The Story of th
(Non-rig

A Study of One of America's Lesser

Known Defense Weapons

Hugh Allen

Alpha Editions

This edition published in 2024

ISBN : 9789362999214

Design and Setting By
Alpha Editions
www.alphaedis.com
Email - info@alphaedis.com

As per information held with us this book is in Public Domain.
This book is a reproduction of an important historical work. Alpha Editions uses the best technology to reproduce historical work in the same manner it was first published to preserve its original nature. Any marks or number seen are left intentionally to preserve its true form.

Contents

Foreword ....- 1 -

CHAPTER I German Submarines in American Waters ....- 3 -

CHAPTER II British Airships in the First War ....- 11 -

CHAPTER III American Airships in Two Wars ....- 15 -

CHAPTER IV The Beginnings of Flight ....- 25 -

CHAPTER V Effect on Aeronautics of Post-War Reaction ....- 34 -

CHAPTER VI Airship Improvements Between Wars ....- 40 -

CHAPTER VII Adventures of the Goodyear Fleet ....- 56 -

CHAPTER VIII Results of Fleet Operations ....- 76 -

CHAPTER IX Vulnerability of Airships ....- 84 -

References ....- 92 -

FOREWORD

High admirals of the American fleet faced in 1940 the gravest responsibility in the National Defense the Navy had ever known. Wherever they turned, north, east, south, west, perils lurked. If they swung their binoculars toward Iceland, toward the Caribbean, toward Singapore, Alaska, or the Canal, everywhere waited potential threats against our American way of life, which they must meet with ships and men, with guns and stout hearts. This was not merely national defense, perhaps not even hemisphere defense, it was World War.

Surveying their gigantic task, and moving swiftly to meet it, they found a place in their program for half forgotten craft, long over-shadowed by other arms of the fleet, the non-rigid airship, sometimes called a dirigible, but more often a "blimp."

Couldn't the airship be used as a watchdog along the coast, against enemy submarines, in discovering enemy mines—relieve for sterner tasks the destroyers and other craft now wallowing their innards out in those restless shallow waters? Great Britain and France had used airships effectively in this service over the English Channel during the last war.

The areas within their patrol range, a hundred or 200 miles out to sea, within the 100 fathom curve, was a vital one. There steamship lanes converge, great harbors lie, coastwise merchantmen cruise, there is the greatest concentration of military and commercial shipping.

With depth bombs and machine guns the blimps might strike a stout blow of their own, even if they weren't rated as combat craft. At least they could sound the alarm, call out reinforcements from swift moving shore-based craft, keep the intruder under surveillance. After all the main thing was to find the submarines in those endless miles of water. And in this field the very slowness of the airship, as compared to the airplane, would be an advantage, permit a more thorough search of the ocean's surface, while its speed as compared to any man-of-war, would enable it to cover more ground within a given 24 hours.

So on the Navy's recommendation Congress in 1940 approved the building of the airship fleet up to substantial proportions, together with bases from which they might operate along the Atlantic and Pacific coasts. That program is now being put into effect and the Goodyear company which had built most of the airships used in the first World War, began again to build ships.

The story of the great rigid airships, the Los Angeles, the Akron, Macon and Graf Zeppelin is fairly well known. That of the smaller non-rigids is less familiar. The larger airships still hold vital commercial and military promise for the future. However, this book will confine itself to the non-rigid airship, with only enough reference to the larger ships to round out the picture.

Every new vehicle of combat or transport has had to fight its way to acceptance against misunderstanding and lack of understanding. Steamships had to prove themselves against sailing ships. Submarines had an uphill battle to establish themselves. The airplane was long on probation, and now the airship is on trial.

This book will tell something about these ships, cite what is claimed for them and what has been reasonably proved they can do, see what progress has been made in performance, and point out what may be expected from them hereafter—not avoiding the moot question of vulnerability.

Lighter-than-air is older by a century than the heavier-than-air branch of aeronautics. Its history is marked by long research and experiment and continued progress. Like every pioneering development it has had its setbacks. But the sincerity of the effort and solid accomplishment made, entitles the project to thoughtful consideration.

CHAPTER I
GERMAN SUBMARINES IN AMERICAN WATERS

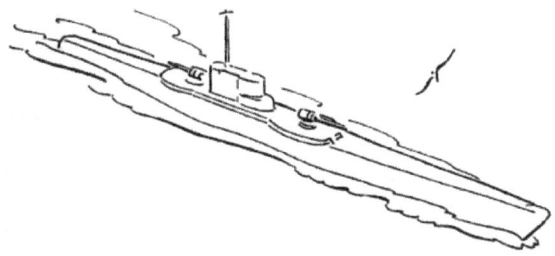

In the last six months of the first World War Germany sent six submarines to America at intervals starting in April, to lay mines along our shipping lanes, attack merchantmen, drive the fishing fleet ashore, try to force this country to call back part of its European fleet for home defense—and in any case to give America, geographically aloof from the war, a taste of what war was like.

These activities were overshadowed at the time by graver events, or hidden by military secrecy. Few people even today know that ships were sunk and men killed by German U-boats within sight of our coast.[1]

It was in no sense an all-out effort. Only a handful of submarines were used. The attack was launched late in the war, in fact one of the six didn't even reach American waters, was called back by news of the Armistice. Submarines of that day had a cruising range of some three months, could spend only three weeks in our coastal waters, used the rest of the time getting over and back.

But in those few weeks these six submarines destroyed exactly 100 ships, of all sizes, types and registry, killed 435 people. Most of the ships were peaceful unarmed merchantmen, coastwise ships from the West Indies and South America, tankers from Galveston, fishing ships heading back from the Grand Banks, supply ships carrying guns and war materials to England, a few stragglers from convoys.

The subs' biggest catch was the USS San Diego, a cruiser, sunk by mine off Fire Island, just outside New York harbor, July 19, 1918, with 1,180 officers and men aboard. Only six lives, fortunately, were lost. The battleship Minnesota, escorted by a destroyer, struck a mine off Fenwick shoals light ship, early in the morning of September 29, but made temporary repairs and limped back into Philadelphia Navy Yard 18 hours later. A fragment of the mine was found imbedded in her frame work.

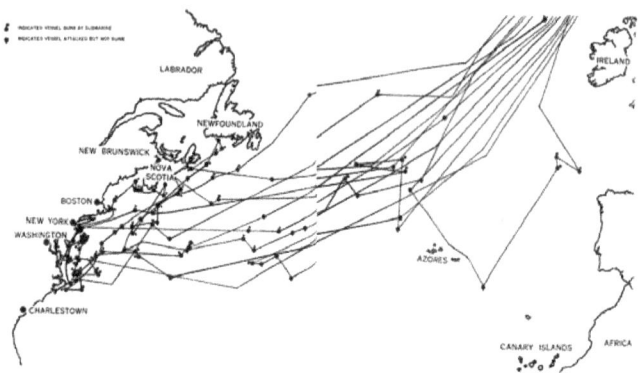

Reproduced from U.S. Navy map showing track of submarines operating in American waters during last few months of first World War.

Mines were laid at strategic points. One field, with its mines 500 to 1,000 yards apart was laid off Cape Hatteras, one at the mouth of Chesapeake Bay, one across Delaware Bay, two in between these key inlets, another off Barnegat, and the last off Fire Island. Some of the mines drifted ashore, others were found and destroyed—the last ones not till the following January. But mines accounted for six of the ships lost.

One of the submarines, the U-117, built as a mine layer, planted 46 of the 58 mines laid along our shores; four others were merchant subs of the Deutschland type, including the Deutschland itself, which had twice previously visited this country on ostensibly friendly missions.

Though the subs encountered a few victims on the way over or back, most of the ships were destroyed in the shallower waters within 200 miles of the American and Canadian coast. The fishing was better close in.

Naval Intelligence knew, through Admiral Sims' office in London, just when each submarine left Kiel, what its probable destination was, and its approximate arrival date. The Navy could not broadcast this information, lest U-boat captains learn they were expected, but took appropriate defense measures. Even so, each submarine traveled directly to its destination, carried out its mission.

U-boats operated almost with immunity from Newfoundland to the Virginia capes. Twice American men of war passed over submerging craft so close as almost to ram them. The U-151 worked at cutting cables for three days, near enough to New York City that the crew could see the lights of Broadway at night. The U-115, lying off the Virginia capes, came to the surface one afternoon just in time for its periscope to disclose a cruiser, two destroyers and a Navy tug a mile away, peacefully returning from routine target practice, entirely unaware that the U-boat was lurking in the vicinity.

The submarines got a poor press that summer, not only for reasons of military secrecy, but because more stirring news held the attention of the public. The AEF was beginning to see action in France.

Still headlines flashed occasionally as censorship was raised, or survivors brought in stories. From the Philadelphia Evening Bulletin during this period:

"Hun U-boats Raid New Jersey coast—Schooner Edward H. Cole Attacked by two Submarines, Destroyed—Two Attacked Off New England—Atlantic Ports Closed"—and the story, under New York date line: "Germany has carried her unrestricted submarine warfare to this side of the ocean—at least five vessels sunk—submarine chasers ordered out from Cape May—Coast Guard stations on special lookout—marine insurance companies announce sharp increase in rates."

News Flash—"Wireless report from passenger steamer Carolina says she is under attack"—The Carolina is sunk, 300 survivors are landed at Barnegat Bay, 19 at Lewes Del., 30 at Atlantic City, others picked up in open boats.

On this map of actual ship sinkings and mine layings in 1918 is superimposed a sketch of the area which a handful of modern patrol blimps might cover.

Then: "Navy mine sweepers sent out to destroy mines and floating torpedoes which had missed target—tanker Herbert L. Pratt strikes mine in shallow water on maiden voyage—War Department asks Congress for $10,000,000 to set up balloon and plane stations along the coast to combat sub menace—British steamer Harpathian torpedoed off Virginia capes—American vessel, name withheld, puts back to 'an Atlantic port' after being chased by U-boat."

The record continues: "San Diego sunk by mine—tug and four barges sunk—British freighter attacked—sub sends landing crew on board lumber schooner off Maine coast, set her afire—Steamer Merak sunk off Hatteras—tanker torpedoed off Barnegat Bay, beaches blanketed with oil—Norwegian steamer Vinland—British steamer Peniston and Swedish steamer Sydland off Nantucket—nine U. S. fishing vessels off Massachusetts coast—British tanker Mirlo—U.S. Schooner

Dorothy Barrett—tanker Frederick R. Kellogg" and so on and on.

Events of the time and since have swept these happenings out of the minds of most Americans—even if they knew of it at the time. But somewhere, half forgotten in Naval files, is an official report, painstakingly compiled after the war, from ship logs, from stories by merchant captains and crews, even by officers of surrendered German submarines, to make up as complete a record as possible of one of the amazing operations of the war—and one whose magnitude, in territory covered and damage done, few suspected, even within the Navy, at the time.

Only two subs had so much as a brush with American ships. The transport von Steuben, former German liner, proceeding to the rescue of men in life boats from a merchant ship, dropped depth bombs which the U-boat escaped by diving to 83 meters, lying low till the enemy had gone.

Closer call had the U-140, largest and most modern of the fleet, which after sinking several ships off Diamond Shoals, including the light ship itself, almost caught a tartar when the Brazilian passenger liner, Uberabe, zigzagging furiously to escape, sent out S.O.S. messages which brought four U.S. destroyers hurrying to the rescue. Nearest was the USS Stringham, which proceeding under full speed, using the Uberabe as a screen, charged on the U-boat, dropped 15 depth charges when the U-boat dived, timed to explode at different levels.

Training exercises with U. S. submarines have taught airship captains much about the habits, movements and characteristics of the underseas craft. (U. S. Navy photo).

The year before America got into the last war the German submarine U-51 sank a half dozen merchant ships off Nantucket Island then proceeded into Newport. (U. S. Navy photo)

Navy airships in practice patrols identify, as to class and nationality, all surface ships in their area, learn to recognize the silhouette of a submarine from afar. (U. S. Navy photo)

The U-boat captain, one of the best in the German navy, drove his craft at a sharp angle to 400 feet. One charge exploding underneath the sub turned it stern upward till it stood almost perpendicular. He managed to level out finally at 415 feet, lay there as long as he dared, finally reached the surface. His ship was so badly crippled it had to abandon its mission and set out for home—though it sunk a couple more ships in the mid-Atlantic on the way back.

The only U-boat casualty was the U-156 which after getting 34 victims in American waters, getting eight in one day, was itself sunk by mines—but off Faroe Island as it was almost home.

This then is the story of submarine operations in U. S. waters in 1918—a half hearted effort of short duration started late in the day—but which destroyed 100 ships, totalling 200,000 tons, most of them close to our shores.

No one could doubt but that in the event of another war submarines would be used again, and in more vigorous fashion. The American fleet might easily keep major enemy ships at a safe distance, and bombing attack from any part of Europe or over the Pacific would have little military value. But certainly submarines would find their way past the screen of Navy craft, bob up off American harbors, again to lay mines in the path of coastwise steamers, deliver hit-and-run attack by torpedo and gunfire at American craft.

We could be equally sure that these ugly motorized sharks, churning the muddy sub-surface waters, would not be satisfied to attack merchantmen only, would be looking for bigger prey.

On the map showing the operations of German submarines in 1918 let us superimpose, as an example, the patrol area which two blimps, basing at Boston, Lakehurst, Cape May and Norfolk might effectively cover in a 12 hour period.

A patrol area of 2,000 square miles per ship is conservative. It assumes the ship flying at no faster than 35 knots, having visibility of five miles in all directions. As a matter of fact, allowing a little more than 40 knots speed—and the airship cruises considerably faster than that—we might say that a modern blimp could patrol an area 10 miles wide and 500 miles long in the 12 hours, or an area of 5,000 square miles. But by criss-crossing back and forth in accordance with a progressive plan, an area of 2,000 square miles could be made reasonably secure—except under extremely adverse conditions of visibility.

Laying these patrol areas down over the map of submarine operations of 1918 it is apparent that such patrols would cover much of the territory where ship sinkings were achieved, cover all of the areas where mines were laid.

With blimps operating from such bases, in addition to the patrols being executed by other naval craft, we might conclude that no submarine could venture within 100 miles of the American coast during daylight hours without considerable risk of detection, and that blimps should be able to make contribution to the safety of coastwise shipping and harbor cities.The patrol areas assigned to the blimps would have their flanks exposed, but airship patrol would be co-ordinated with that of airplanes and surface craft, guarding the areas farther out.

That this conclusion is reasonable is indicated by the fact that from 1939 on, Lakehurst Naval Air Station, under command of Commander G. H. Mills had been doing just this, patroling areas all the way from Nantucket to Cape Hatteras.

CHAPTER II
BRITISH AIRSHIPS IN THE FIRST WAR

Germany entered the first World War with high expectations as to one, perhaps two of its new weapons of war. Its submarines might offset Britain's superiority at sea, and certainly the Zeppelins, which had proved themselves in four years of commercial flying, would be able to cross the English Channel and carry the war to the island which had seen no invasion since William the Conqueror.

No nation except Germany had Zeppelins. And as the German people began to feel the pinch of the blockade, cutting their life line of food and supplies, they brought increasing public pressure on High Command to use these weapons to punish England.

Later commentators have speculated as to whether, if Germany had held its fire, waited till it could assemble an overpowering force of Zeppelins and submarines and stage a joint attack, it might not have been able to force a quick decision.

But the Zeppelins were sent over a few at a time, as fast as they could be built, and England was given time to devise defenses. These were chiefly higher altitude airplanes, farther ranging anti-aircraft guns, sky piercing searchlights, which combined to force the invaders to fly continuously higher as the war wore on, as high as 25,000 feet at times, with corresponding sacrifice of bombing accuracy. And when machine guns, synchronized with the propellers, were mounted in airplane cockpits, and began to spit inflammable bullets into the hydrogen filled bags and send them down in flames, the duel took on more even terms.

Less spectacularly the Zeppelins were used on a wide scale as reconnaissance and scouting craft, which flying fast and far were given credit on more than one occasion for saving German Naval squadrons from being cut off by superior Allied forces, were acknowledged even by the British to have played an important part in the Battle of Jutland.

It is a little hard to realize today that whatever air battles were waged over water in the last war were conducted chiefly by lighter-than-air craft. Planes staged spectacular battles along the Allied lines in France, but lack of range and carrying capacity forced them to leave sea battles to the airship. As a measure of that situation, the great hangars at Friedrichshafen, spawning ground of the Zeppelins, one of the outstanding targets in all Europe if England were to draw the dirigible's fangs, lay hardly more than a hundred miles from the French borders, but even that distance was too great for effective attack.

While these greater events were taking place, British airships, smaller in size, less spectacular, were playing no small part in repelling Germany's other threat, the submarine.

BLIMPS USED TO SEARCH FOR U-BOATS

Navy opinion around the world was skeptical at the beginning of the War as to whether submarines would ever be practical. There were mechanical troubles, accidents, usually costly. Even Germany, prior to 1914, used to send an escort of warships along to convoy its subs to their station—then send out for them afterward to bring them home again.

But the war was only a few weeks old when the captain of the U-9, cruising down the Dutch coast, discovered that his gyro compass was off, and when he got his bearings saw that he was 50 miles off course. He wasted no breath, however, on many-syllabled German swear words, for off on his southern horizon were the masts of three British ships. He dived, came up alongside, and in 30 minutes, single handed, with well directed torpedoes, had sunk in turn HMS Aboukir, Hogue and Cressy.

The morning of September 22, 1914, marked the beginning of a new era in Naval warfare. The warring nations grew furiously busy building their own U-boats and devising defenses against the enemy's. Among these defenses was the non-rigid airship.

These two vehicles, so widely different, have much in common. If we may be technical for a minute we may say that the airship and the submarine are both buoyant bodies, completely immersed and floating in a medium—air and water respectively—of changing pressures, that each uses dual sets of steering gear and rudders to control direction and altitude. And further, that the airship in 1941 faces the same division of opinion as the submarine faced in 1914, as to whether, particularly with rigid airships, it will ever be widely used and accepted.

In any event in 1914 there was an urgent and immediate job to be done.

Indicator nets and high explosive mines might give some protection to harbors, might be stretched across steamship lanes and planted around the hiding places of the submarines, if those could be discovered. But troop ships and munition ships and food ships must be dispatched without interruption across the tricky waters of the English Channel to France, and for this purpose convoy escorts were devised, with camouflaged warships zigzagging alongside, while high aloft in lookout stations men with binoculars strained their eyes, searching the waters, ahead, astern, alongside, their search lingering long over every bit of floating wreckage—and there was a lot of it—to make sure it was not a periscope.

These lookouts aboard ship quickly had a new ally in the air. As the submarine menace grew, binoculars began to flash too from the fuselages of bobbing blimps overhead. At a few hundred or perhaps a thousand feet elevation they could see deep below the surface, and quickly learned to recognize at considerable distance the tell-tale trail of bubbles or feathered waters or smear of oil which denoted the enemy's presence, might even pick out the shadowy form of the submerged craft itself.

The value of the airship in convoy was that it could fly slowly, could throttle down its motors and march in step with its charges, cruise ahead, alongside, behind. The very speed of its sister craft, the airplane, handicapped its use in this field.

This characteristic of the blimp was even more useful in hunting U-boat nests. The blimp could head into the wind, with its motors barely turning over, hover for hours at zero speed over suspect areas. It could fly at low altitudes, follow even slender clues. Seagulls following a periscope sometimes gave highly useful information. An orange crate moving against the tide attracted the attention of one alert pilot, for the crate concealed a periscope, and the blimp dropped bombs—successfully.

When a blimp discovered a submarine, it would give chase. With its 50 knots of reserve speed it was faster than any warship, much faster than the poky wartime submarine, which could do only 10 or 12 knots on the surface, much less than half that when submerged. If it was lucky the airship might drop a bomb alongside before the sub got away.

And run for cover the submarine always did. It wanted no argument with a ship which could see it under water, could outrun it, and might plunk a bomb alongside before its presence was even suspected.

Airships did get their subs during the war. The records, always incomplete in the case of submarines, whose casualties were invisible, show that British blimps sighted 49 U-boats, led to the destruction of 27. But their greater usefulness lay in the fact that their mere presence sent the underseas craft scuttling for submerged safety.

Between June, 1917, and the end of the war British blimps flew 1,500,000 miles, nearly as many as the Zeppelins. A French Commission made an exhaustive study of dirigible operations after the war, and the late Rear Admiral W. A. Moffett quoted from its reports in summarizing lighter-than-air lessons taken from the war, when he told the Naval Affairs Committee of the House of Representatives that "as far as they could learn, no steamer was ever molested by submarines when escorted by a non-rigid airship."

France and Italy had long coast lines, used the blimps extensively along the Bay of Biscay, the Mediterranean and the Adriatic, but England found still greater use for them because it was an island. So blimp scouts played a singularly useful role from Land's End to the Orkneys, stood watch at the mouth of the Firth of Forth, the Solway, the Humber, and the Thames.

CHAPTER III
AMERICAN AIRSHIPS IN TWO WARS

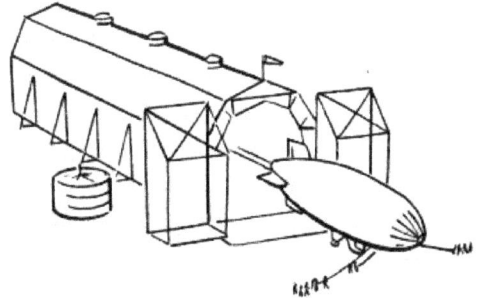

Compared to British and French airships, American dirigibles made a less impressive record during the first war.

This for the reasons that there were few enemy activities in our waters until the very end, and that there were few American airships to oppose them. Virtually the entire airship organization had to be created after we got into the war.

Naval attachés abroad had been watching blimp operations over the English channel, and on the basis of rather meager information which they furnished, Navy designers were working on plans, when the Secretary of the Navy, in February 1917, 60 days before the declaration of war, ordered 16 blimps started at once.

Nine of these were to be built by Goodyear which had at least given some study to the principles, had built a few balloons, one of which, flown by its engineers out of Paris, had won the James Gordon Bennett Cup Race.

No one in this country, however, knew much about building airships, and less about flying them after they were built. Operating bases would have to be built and the very construction plants as well. The first Goodyear airship under the Navy order was completed before the airship dock (hangar) at Wingfoot Lake was ready, and the ship had to be erected in Chicago and flown in.

The engineers who built it, Upson and Preston, made their first airship flight in delivering the ship to Akron, using theoretical principles applied in the international balloon race the year before, to make up for their lack of practical experience.

Those first ships were small, slow, lacked range, uncovered many shortcomings. Flight training was done under adverse circumstances. Men had to teach themselves to fly airships, then teach others to fly them.

The pilots were chiefly engineering students from the colleges, with a sprinkling of Navy officers. They had to take their advanced training abroad at British and French bases, because there were no facilities here, and in fact did most of their flying abroad. By the end of the war American pilots were manning three British airship bases and had taken over practically all the French operations, including the large base at Paimboeuf, across the Loire from St. Nazaire, on the French coast.

So the war was well along before American bases were set up and manned. These were at Chatham, Mass., at Montauk and Rockaway, N. Y., at Cape May, Norfolk and Key West. Like the airplane patrols the blimps saw little action, though they had an advantage in that they could stay out all day, while the short range planes of 1917-18 had to come back every few hours to refuel.

A patrol airship at Chatham, Mass. missed its chance in that it was adrift at sea with engine trouble when the German U-156 slipped into the harbor at nearby Orleans and wiped out some fishing boats—though it might have done no better than the first plane which reached the scene, whose few bombs did not explode.

The blimp patrols, however, uncovered one other type of activity. More than once they spotted suspicious looking craft emerging under cover of fog, from remote coves and inlets along the Long Island coast, fishing boats and barges with improvised power plant and curious looking paraphernalia on deck. Keeping the stranger in sight the blimp summoned armored craft from shore which sent boarding crews on, found mines destined for the New York steamship lanes.

A more important result of the blimp operations was the improvements in design which were found, particularly in the "C" type ship, brought out in 1918, of which 20 were built. They had much better performance in range, power, could make 60 miles speed, were faster than any airships except the Zeppelins. Navy officers and crews came to have high respect for them.

Here's the gallant C-5, which with a bit of luck would have been the first aircraft to cross the Atlantic. (U. S. Navy photo)

Wingfoot Lake, Akron, was a busy place during the first war, as the spawning ground of scores of blimps, hundreds of training and observation balloons.

"Finger patches" of rope ends raveled out and cemented to the outside of the bag were used in 1918 to support the weight of the gondola—an improvised airplane fuselage.

During most of the period between World wars the Navy had only a few J-type ships, but used them effectively in training and experimental work. (U. S. Navy photo)

Which led to one of the interesting aeronautic adventure stories of the period. It happened just after the Armistice.

Men had come out of the war with imaginations afire over the possibilities of aircraft. One challenge lay open—the Atlantic—no one had flown it.

In the breathing spell brought by the Armistice, the British were preparing their new Zeppelin R-34 for the crossing; two English planes were being shipped to Newfoundland to try to fly back; the U. S. Navy had a seaplane crossing in prospect. There was even a German plan. A new Zeppelin had just been finished at Friedrichshafen when the Armistice was signed, and the crew planned to fly it to America as a demonstration—but authorities got wind of it and blocked the venture.

But of all the Atlantic crossings about which men were dreaming in early 1919, none is more interesting than the one projected for the little blimps.

The C-5, newest of the non-rigid airships built for the Navy, was stationed at Montauk, and there one night a group of officers sat intensively studying charts and weather maps. St. John's, Newfoundland, 1,400 miles away, would be the first leg of the trip. It was easily within the cruising radius of the ship, particularly if they got helping winds, which they should if the time was carefully picked. From there to Ireland was another 1850 miles, also within range with the prevailing westerly winds.

Permission was asked from Washington, and the Navy flashed back its approval and its blessing, assigned five experienced officers to the project: Lieut. Comdr. Coil, Lieuts. Lawrence, Little, Preston, and Peck. The USS Chicago was sent ahead to St. John's to stand by and give any help needed.

Shortly after sunrise on May 15, 1919, motors were warmed up and the ship shoved off from the tip of Long Island with six men aboard headed for Newfoundland. At 7 o'clock the next morning they circled over the deck of the *Chicago*, dropped their handling lines to the waiting ground crew on a rocky point at St. John's. The first leg had been made in a little more than 24 hours, at an average speed of nearly 60 miles per hour.

The morning was clear and comparatively calm. Coil and Lawrence went aboard the *Chicago* to catch a little sleep before the final hop over the ocean. The others saw to re-fueling the C-5, stowing provisions aboard, topping off a bit of hydrogen from the cylinders alongside. Mechanics swarmed over the motors. All was well.

But about 10 o'clock gusts began to sweep down from Hudson's Bay, dragging the ground crew over the rocks. There were no mooring masts in those days. A modern mast would have saved the ship. More sailors were put on the lines and word sent to Coil and Lawrence. If the ground crew could hold the ship till the pilots could get aboard and cut loose, the storm would give them a flying start over the Atlantic.

But the wind blew steadily stronger as the commander was hurrying ashore. It reached gale force, hurricane force, 40 knots, 60 knots in gusts, varying in direction crazily around a 60-degree arc. It picked the ship up and slammed it down, damaging the fuselage, breaking a propeller. Little and Peck climbed aboard to pull the rip panel and let the gas out. After the storm passed, they could cement the panel back in, reinflate the bag and go on.

But the fates were against them. The cord leading to the rip panel broke. Desperately, the two men started climbing up the suspension cables to the gas bag with knives, planning to rip the panel out by hand. But a tremendous gust caught the ship, lifted it up. Seeing the danger to the crew, Peck shouted to them to let go, and he and Little dropped over the side. Little broke an ankle.

The ship surged upward, crewless, set off like another "Flying Dutchman" across the Atlantic, was never seen again.

Just three days later Hawker and Grieve set out from St. John's, landed in the ocean. Alcock and Brown cut loose their landing gear a month later and landed in Ireland. One of the three Navy seaplanes, the NC-4, reached Europe on May 31 and the British dirigible R-34 set out on July 2 for its successful round trip to Mitchel Field.

But for a trick of fate and the lack of equipment available today, a blimp would have been first to get across.

Many things happened in the airship field between the two wars, but most of them affected non-rigid airships only indirectly, as the Navy was primarily concerned with the larger rigids.

The loss of the Hindenburg by hydrogen fire (which American helium would have prevented), coming on the heels of tragic setbacks in this country was enough to dismay anyone except

Commander C. E. Rosendahl and his stouthearted associates at Lakehurst Naval Air Station.

They didn't give up. Setbacks were inevitable to progress. Count Zeppelin had built and lost five rigid airships prior to 1909, but he went on to build ships which were flown successfully in war and peace. If the Germans, using hydrogen, could do this, Americans, with helium, should not find it impossible, Lakehurst reasoned. And if they had no rigid airships to fly and no immediate likelihood of getting any they would use blimps.

The Navy was more familiar than the public with what the British and French airships had accomplished in the first war. Studying, as all Navy officers were doing in that period, the various possibilities of attack and defense, in case the war then threatening Europe should sweep across the Atlantic, they came to the conclusion that the coast line of America was no more remote from German submarines in 1938 than the coast of England was in 1914.

The airplane had improved vastly in speed, range, and striking power, and their very multiplicity had ruled out the blimps over the English channel, even if helium was available, but those conclusions did not hold along the American coast.

The heroic part played by Allied blimps was a part of the legend of the airship service. Nothing new developed in war had subtracted anything from the ability of American airships to do in this war what British non-rigids had done in the last. Commander J. L. Kenworthy and after him Commander G. H. Mills as commanding officer at Lakehurst turned to non-rigids.

Under Mills was instituted, quietly, unostentatiously, with what ships he had, a series of practice patrols to determine the usefulness of airships in this field, to discover and perfect technique, and to train officers and men.

Lakehurst had a curious conglomeration of airships to start with. There were two J ships of immediate post-war type, with open cockpits, 210,000 cubic feet capacity; two TC ships, inherited from the Army, of more modern design, and larger size; the ZMC2, an experimental job built to study the use of a metal cover, and about to be scrapped after nine years of existence; the L-1, the same size as the Goodyear ships, 123,000 cubic feet, the first modern training ship, which would be joined later by the L-2 and L-3; the G-1, a larger trainer of Goodyear Defender size, useful for group instruction, and the 320,000 cubic foot K-1, which had been built for experiments in the use of fuel gas. Only the K-2, prototype of the 416,000 cubic foot patrol ships later ordered could be called a modern airship, though the Army dirigibles also had good cruising radius.

Yet with this curious assortment of airships of various sizes, types and ages, the Navy carried on practice patrols covering the areas between Montauk and the Virginia Capes, flying day after day, built an impressive accumulation of flight data, missed very few days on account of weather, made it a point not to miss a rendezvous with the surface fleet. More than any one thing it was this demonstration, over an 18-month period, which led to the revival of an airship program in this country, the ordering of ships and land bases.

Let us see what a blimp patrol is like. The airship can fly up to 65 knots or better, but this is no speed flight. The motors are throttled down to 40 knots, so that the crew may see clearly, take its time, study the moving surface underneath, scrutinize every trace of oil smear on the surface, alert for the tell-tale "feather" of the submarine's wake, air bubbles, a phosphorescent glow at night, for even a bit of debris which might conceal a periscope.

A school of whales, a lone hammerhead shark on the surface or submerged stirs the interest of the patrol, offers a tempting live target for the bombs,—light charges with little more powder than a shot gun shell uses. Now a ship records a direct hit on a shark's back 500 feet below. He shakes his head, dives to escape this unseen enemy aloft. The airship gives chase, follows the moving shadow below, so strikingly resembling a submarine, finds the practice useful.

Of the crew of eight, everyone on the airship is on watch, with an observation tower open on all sides, without interference of wings, as in the airplane. Compared with surface craft, the airship can patrol more area in a dawn to dusk patrol because of its speed and its wide range of unbroken observation.

The submarine is more efficient in relatively shallow depths, but airships have spotted the silhouetted shadow of U-boats in clear water as deep as 70 feet below the surface. The submarine will attempt to maneuver within a mile of its target to launch its torpedoes effectively. But even at a mile away the ten inches of periscope which projects above the surface is difficult for other craft to detect,—either for a cruiser at sea level, or an airplane, flying at relatively high speed, a threat either may miss.

Airship crews are at action stations even during peace time, on the alert against the appearance of strange craft. They identify each passing ship through binoculars, by voice or radio, taking no chances that attack without warning by a seeming peaceful ship might not be a declaration of war. As many as 40 or 50 ships may be encountered and identified in a day's patrol. The airships are off at sun-up, back at sundown, unless on more extended reconnaissance, move quietly into the big dock.

Patrol is tedious work. Occasionally there is a break in the routine. Lt. Boyd has been assigned command of the big TC-14 for the next day's patrol. He is up late studying the curious tracks he is to follow in coordination with the other airships. At midnight however the radio brings startling word. An airplane leaving Norfolk with a crew of ten for Newport, is reported missing. Nearby destroyers, airplanes, airships, are ordered out as a searching party. The TC-14, having the longest cruising radius, 52 hours without refueling, is sent off at once, with a senior officer, Lt. Trotter, in charge. Men's lives may be at stake.

By daylight, the TC-14 has flown over the entire northern half of the plane's track and back, watching intently for distress signals or flares or any sign of the distressed plane. Three miles north of Hog Island light outside Norfolk, the ship encounters fog extending clear to the water. Search of this area is hopeless and the ship scouts the edges, waits for the fog to burn off. At noon as it lifts, pieces of wreckage are spotted at the very area which it had hidden, and which the TC-14 had flown over five hours earlier.

The airship cruised around, hoping that some bit of wreckage might support a survivor, finally returned to its station after 20 hours, during which time it had covered 1,000 miles, intensively in parallel courses 20 miles wide. Had the luckless plane or any of its crew been able to send up flares anywhere within an area of 20,000 square miles of water, the airship could have come up alongside and effected a rescue in a matter of minutes.

In the meantime, Lt. Boyd, originally assigned to TC-14, was up at dawn only to learn of the change in plans. He was assigned to pilot the smaller G-1 trainer to New London, keep a sharp lookout enroute for the missing plane, then work with the destroyers on torpedo exercises. The G-1 had no galley aboard and in the rush the matter of food for an 18-hour cruise was somehow overlooked, and Boyd and his crew set off with only a couple of sardine sandwiches apiece and a pot of coffee, which quickly grew cold.

Late in the afternoon, seeing his crew growing hungrier and hungrier,—for airshipping is excellent for the appetite,—Boyd had an idea. He radioed the Destroyer Division Commander: "After last torpedo recovered, would you be able to furnish us with some hot coffee and a loaf of bread, if we lower a container on a 200-foot line across your after deck?"

Never in naval history had an airship borrowed chow from a surface craft. But the answer came promptly. "Affirmative. Do you wish cream and sugar?"

There was nothing in the books giving the procedure for borrowing a meal from the air, but the crew rigged up a line from a target sleeve reel, fastened a hook with a quick release at the end, attached a monkey wrench to weight it down, stood by for the word to come alongside.

Then while the crews of three destroyers watched, the G-1 swung slowly over the destroyer's deck. One sailor caught the line held it while a second filled the coffee pot, and a third attached a load of sandwiches. Then the airship sailors hauled away, radioed their thanks, set off for the 200 mile trip back to Lakehurst, while hundreds of sailors below waved their white caps and cheered, a little inter-ship courtesy between sky and sea which all hands will long remember.

CHAPTER IV
THE BEGINNINGS OF FLIGHT

In the spring of 1783, as the American Revolution was nearing a successful conclusion, two brothers named Montgolfier sitting before a fire at a little town in France found themselves wondering why smoke went up into the air.

That was just as foolish as Newton wondering why an apple, detached from the tree, fell down. Smoke had always gone up and apples had always come down. That was all there was to it.

But when men wonder momentous events may be in the making. In these instances epochal discoveries resulted: the law of gravitation and the possibility of human flight.

The legends of Icarus and the narrative of Darius Green are symbols of the long ambition of earth-bound men, even before the days of recorded history, to leave the earth and soar into the air. The Montgolfiers had found the key.

But a hundred years would pass before the discovery would be put to use. It was in 1903 that another pair of brothers, the Wrights, made their first flight from Kill Devil Hill in North Carolina. The first Zeppelin took off from the shores of Lake Constance in 1900.

The Montgolfiers wasted no time testing out their conclusion that smoke rose because it was lighter than the air. They built a great paper bag 35 feet high, hung a brazier of burning charcoal under it, and off it went. Annonnay is a small town but the story of that miracle spread far and wide. The Academy of Science invited them to the capital to repeat the experiment.

But while they were building a new bag a French physicist, Prof. J. A. C. Charles, stole a march on them. He knew that hydrogen was also lighter than air, so constructed a bag of silk, inflated it with hydrogen, sent it aloft before the Montgolfiers were ready.

Still the countrymen were not to lose their hour of glory. Merely to repeat what had already been done was not enough. Their balloon was to be flown from the grounds of the Palace of Versailles, before the king and court and all the great folk of Paris, with half the people of the city craning their necks to watch it pass over. So they loaded aboard a basket containing a sheep, a duck and a rooster, and these three became aircraft's first passengers.

When the U. S. Army Air Corps years later sought an appropriate insignia for its lighter-than-air division, it could think of nothing more fitting than a design which included a rooster, a duck and a sheep.

Everyone was ready for the next step. A French judge had the solution. He offered the choice to several prisoners awaiting execution—a balloon flight or the guillotine. Two volunteered, felt they had at least a chance with the balloon, whereas the guillotine was distressingly final. They had nothing to lose. That word rang through Paris. A young gallant named De Rozier objected.

"The chance might succeed," he said. "The honor of being the first man to fly should not go to a convict, but to a gentleman of France. I offer my life."

Even the king protested at this needless risk, but De Rozier took off the following month, flew half way over Paris, landed safely. This happened on Nov. 21, 1783.

Among the witnesses to these experiments was Benjamin Franklin, the American ambassador, himself a scientist of no small renown. He predicted great things for aeronautics.

"But of what use is a balloon?" asked a practical-minded friend.

"Of what use," replied the American, "is a baby?"

A little later, on January 7, 1785, Jean Pierre Francois Blanchard, a Frenchman, and Dr. John Jeffries, an American, practicing medicine in England, inflated a balloon, took off from the cliffs of Dover at one o'clock in the afternoon, arrived safely in Calais three hours later.

Santos Dumont startled Paris in 1910, when he let an American girl fly one of his airships over the city. To descend she threw her weight forward, to climb she moved back a step.

A dramatic meeting of two rivals for the honor of making the first Atlantic crossing. The Navy's NC flying boats and the non-rigid C-5, photographed shortly before their take-off.

Blimps too may use masts aboard surface ships as anchorage point on long cruises, as the U.S.S. Los Angeles successfully demonstrated when moored to the U.S.S. Patoka. (U. S. Navy photo)

The Army's TC-7 demonstrates the first airplane pick-up at Dayton. Army pilots found that at flying speed the plane weighed nothing, was sustained by dynamic forces. (U. S. Army photo)

Flight was here, though it would be a long time becoming practical. Dr. Charles and many others contributed, even at that early day. Knowing that hydrogen expanded as the air pressure grew less, at higher altitudes, Charles devised a valve at the top of the balloon, so that the surplus gas could be released, not burst the balloon. He devised a net from which the basket could be suspended, distributing its load over the entire bag.

The drag rope was evolved, an ingenious device to stabilize the balloon's flight in unstable air. If the balloon tended to rise it would have to carry the entire weight of the rope. If it grew sluggish and drifted low, it had less weight to carry, as much of the rope now lay on the ground. These ballooning principles, early found, are still in use. But the "dirigible" balloon, or airship must wait for light weight, dependable motors, despite the hundreds of ingenious experiments made by men over a full century.

Since this is an airship story, we should first make clear the difference between the airship and the airplane.

The French hit on an apt phrase to distinguish them, dividing aircraft into those which are lighter than the air, such as airships, and those which are heavier than the air, like airplanes.

Airships are literally lighter than air. So are all free balloons, used for training and racing, and all anchored balloons, such as the observation balloon widely used in the last war and the barrage balloons of the present war.

The airship goes up and stays up because the buoyancy given by its lifting gas makes it actually lighter than the air it displaces, and even with the load of motors, fuel, equipment and passengers, must still use ballast to hold it in equilibrium.

The airplane, on the other hand, is heavier than the air. Even the lightest plane can stay up only if it is moving fast enough to get a lifting effect from the movement of air along the wings, similar to that which makes a kite stay up. A kite may be flown in calm weather only if the one who holds the cord keeps running. On a windy day, the kite may be anchored on the ground, and the movement of the wind alone will have sufficient lifting effect. So powerful are these air forces that a plane weighing 20 tons may climb to an altitude of 10,000 feet if its speed is great enough, and its area of wing surface broad enough to produce this kiting effect.

But an airplane can remain aloft only as long as it is moving faster than a certain minimum speed. Cut the motors, or even throttle down below this stalling speed, and the plane will start earthward.

The airship needs its motors only to propel it forward. It can cut its speed, even stop its engines, and nothing happens. It retains its buoyancy, continues to float. The airplane's lift is dynamic, that of the airship is static.

The airship has some dynamic lift, also, because its horizontal fins or rudders, and the body of the airship have some kiting effect in flight. The blimp pilot, starting on a long trip, will fill up his tanks with all the fuel the ship can lift statically, then take on another 2,000 pounds, taxi across the airport till he gets flying speed and so get under way with many more miles added to his cruising speed.

This dynamic lift however, while useful in certain operations is still incidental. Primarily the airship gets its lift from the fact that the gas in the envelope is much lighter than the air.

Hydrogen is only one-fifteenth the weight of air, helium, the non-inflammable American gas, is a little heavier, about one-seventh. The practical lift is 68 pounds to the thousand cubic feet of hydrogen, 63 pounds in the case of helium.

Lighter-than-air ships are of three classes, rigid, semi-rigid and non-rigid. The rigid airship has a complete metal skeleton, which gives the ship strength and shape. Into the metal frame of the rigid airship are built quarters, shops, communication ways, even engine rooms in the case of the Akron and Macon, with only the control car, fins, and propellers projecting outside the symmetrical hull. The lifting gas is carried in a dozen or more separate gas cells, nested within the bays of the ship.

The non-rigid airship has no such internal support. The bag keeps its taut shape only from the gas and air pressure maintained within. Release the gas and the bag becomes merely a flabby mass of fabric on the hangar floor. Ship crews do not live in the balloon section, but in the control car below.

The British, apt at nicknames, differentiated between the two types of airships by calling them "rigid" and "limp" types, and since an early "Type B" was widely used in the first World War, quickly contracted "B, limp" into the handier word "Blimp."

The third type, semi-rigid, has a metal keel extending the length of the ship, to which control surfaces and the car are attached, and with a metal cone to stiffen the bow section.

The rigid ship is of German origin. Developed by Count Zeppelin, retired army officer, and largely used by that nation during the war of 1914-18, it was taken up after the war started, by the British and Americans, and to a small extent later by France and Italy.

Non-rigid ships were widely used by the British and French, to a less extent by Italy and United States.

The intermediate semi-rigid was largely Italian and French in war use, though United States bought one ship after the war from the Italians, built one itself. The Germans also built smaller Parseval semi-rigids.

The rigid airships are the largest, the non-rigids smallest. The rigid has to be large to hold enough gas to lift its metal frame along with the load of fuel, oil, crew, supplies, passengers and cargo. The blimps can be much smaller.

The Army's first airship, built by Major Tom Baldwin, old time balloonist, had 19,500 cubic feet capacity. Goodyear's pioneer helium ship "Pilgrim" had 51,000 cubic feet. These contrast with the seven million feet capacity of the Hindenburg, and the ten million cubic feet of ships projected for the future.

The following table will show the range of sizes:

Rigid Airships:	Hindenburg (German)	7,070,000 cubic feet
	Akron-Macon (U. S.)	6,500,000 cubic feet
	R-100, 101 (British)	5,000,000 cubic feet
	Graf Zeppelin (German)	3,700,000 cubic feet
	Los Angeles (U. S.)	2,500,000 cubic feet
	R-34 (British)	2,000,000 cubic feet
Semi-Rigids:	Norge (Italian)	670,000 cubic feet
	RS-1 (U. S.)	719,000 cubic feet
Non-Rigids:	Navy K type (Patrol)	416,000 cubic feet
	Navy G type (Advanced Training)	180,000 cubic feet
	Navy L type (Trainer)	123,000 cubic feet

Goodyear (Passenger)	123,000 cubic feet
Pilgrim (Goodyear)	51,000 cubic feet

The Akron and Macon were 785 feet in length, the K type non-rigid, 250 feet long, the Navy "L's" 150 feet long.

Let's cut back now to the Montgolfiers. Progress was disappointingly slow. The simple balloon would only go up and down, and in the direction of the wind. Before it could be practical, men must be able to drive it wherever they liked, make it dirigible, or directable.

Ingenious men, Meusnier, Giffard, Tissandier, Renard, Krebs, many others worked over that problem through the entire nineteenth century. They devised ballonets or air compartments to keep the pressure up. They built airships of cylinder shape, spindle shape, torpedo shape, airships shaped like a cigar, like a string bean, like a whale. But the stumbling block remained, the need of an efficient power plant.

The steam engine was dependable, but once you had installed firebox, boiler and cord wood aboard, there was little if any lift remaining for crew or cargo. Giffard in 1852 built an ingenious small engine using steam but it still weighed 100 pounds per horsepower, drove the ship at a speed of only three miles an hour. Automobile engines today weigh as little as six pounds per horsepower, modern airplane engines one pound per horsepower.

Man experimented with feather-bladed oars, with a screw propeller, turned by hand, using a crew of eight men. Haenlein, German, built a motor that would use the lifting gas from the ship—coal gas or hydrogen. Rennard in 1884 built an electric motor, taking power from a storage battery.

But real progress would have to wait for the discovery of petroleum in Pennsylvania and the invention of the internal combustion engine. When the gasoline engine came in, in the 90's, the dirigible builders saw the long sought key to their problem.

While Count Zeppelin was experimenting with his big ships in Germany, Lebaudy, Juliot, Clement Bayard in France and most conspicuously the young Brazilian, Santos Dumont, were working with the smaller dirigibles. Santos Dumont built 14 airships in the first decade of the century, brought the attention of the world to this project. He won a 100,000 franc prize in 1901 for flying across Paris to circle Eiffel Tower and return to his starting point—and gave the money to the Paris poor.

The Wright Brothers made their historic flight at Kitty Hawk, in 1903, opening a different field of experiment. France pushed both lines of research. After Santos Dumont's dirigible flight, Bleriot started from the little town of Toury in an airplane, flew to the next town and back, a distance of 17 miles, making only two en route stops,—and the town erected a monument to him.

In 1909, Bleriot flew a plane across the English Channel and in the following year the airship Clement Bayard II duplicated the feat, carrying a crew of seven, made the 242 miles to London in six hours.

The year 1910 was a momentous one for all aircraft, with France as the world center. Bleriot and Farman, Frenchmen, Latham, British, the Wrights and Curtiss, Americans, broke records almost daily at a big meet in August that year, while at longer range the French and English dirigibles and the Parsevals of Germany, and still more important the great Zeppelins at Lake Constance droned the news of a new epoch.

A young American engineer, P. W. Litchfield, attended the Paris meet, saw these wonders, made notes. He stopped in Scotland on his way back, bought a machine for spreading rubber on fabric, hired the two men tending it (those men, Ferguson and Aikman, were still at their posts in Akron thirty odd years later), hired two young technical graduates on his return, tied in the fortunes of his struggling company with what he believed was a coming industry.

The next five years would see the nations of the world bending their efforts toward perfecting these vehicles of flight,—little realizing they were building a combat weapon which would revolutionize warfare.

CHAPTER V
EFFECT ON AERONAUTICS OF POST-WAR REACTION

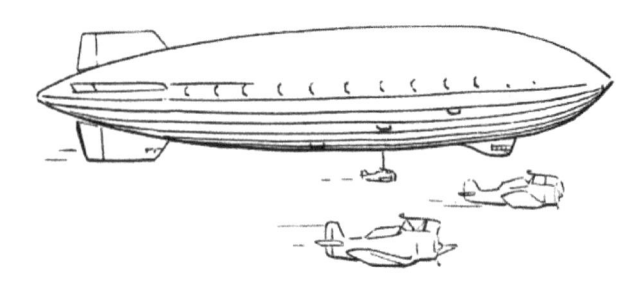

Development of non-rigid airships slowed down after the impetus of the war had spent itself, as was the case in aeronautics generally and in all defense efforts.

With the Armistice of November, 1918, the world was through with war. Men relaxed and reaction set in. There would not be another major war in a hundred years. Well-meaning people everywhere grasped at the straw of universal peace, of negotiated settlement of difficulties between nations, of disarmament of military forces to the point of being little more than an international police force. Germany, the trouble-maker, had been disarmed and handcuffed, would make no more trouble. The world, breathing freely after four years, wanted only to be left alone.

Today with major countries striving feverishly to build guns and navies, it is hard to believe that naïve nations were scrapping ships only a few years ago and pledging themselves to limit future building. No one in the immediate post-war era could believe that men must prepare for another war, an all-out war more terrible and ruthless than men had known,—one which would send flame-spitting machines down from the air and through woods and fields, against which conventional foot soldiers would be as helpless as if they carried bows and arrows. Wishing only to live at peace with other nations, we could conceive no need to make defense preparation against frightfulness.

Congress was divided between "big navy men" and "little navy men," and generals and admirals who brought in programs for expansion or even reasonable maintenance were shouted down. The public was in no mood to listen.

If the usefulness of the Army and Navy was discounted during this period, more so was the rising new Air Force. Few were interested in airplanes, and these chiefly wartime pilots, who sought to keep aviation alive, made a precarious living flying wartime "Jennies" and "Standards" out of cow pastures, carrying passengers at a dollar a head, or how much have you. The word "haywire" came into the language, as they made open-air repairs to wings and fuselage with baling wire.

Lighter-than-air had no Rickenbackers or Richthofens to point to, but got some advantage during this period from the activities of the Shenandoah, completed in 1923, and the Los Angeles, delivered in 1924. These ships could not be regarded as military craft, carried no arms. The Shenandoah was experimental, based on a 1916 design. The Los Angeles was technically a commercial ship, with passenger accommodations built in, could be used only for training.

This grew out of the fact that the Allies planned to order the Zeppelin works at Friedrichshafen torn down but had held up the order long enough for it to turn out one more ship. This last ship would be given to United States in lieu of the Zeppelin this country would have received from Germany, if the airship crews, like those of the surface fleet, had not scuttled their craft after the Armistice, to keep them from falling into enemy hands. The Allies stipulated that the Los Angeles should carry no armament. It took a specific waiver from them for the ship to take part several years later in fleet maneuvers.

Other airship activities in this country were at a minimum. The blimps, little heard of in this country during War I, remained in the background. A joint board of the two services gave the Navy responsibility for developing rigid airships, the Army to take non-rigids and semi-rigids. The Navy maintained a few post-war blimps for training, had little funds except for maintenance.

The Army, having Wright Field to do its engineering and experimental work, fared somewhat better, carried on a training and something of a development program. It built bases at Scott Field, Ill., and Langley Field, Va., ordered one or two non-rigid ships a year, purchased a semi-rigid ship from Italy, ordered another, the RS-1, from Goodyear, operated it successfully.

The Army's non-rigids, however, were overshadowed by the Navy's rigids and even more by its own airplanes, with the result finally that the Chief of the Air Corps, Major General O. O. Westover, a believer in lighter-than-air, an airship as well as airplane pilot, and a former winner of the James Gordon Bennett cup in international balloon racing, told Congress bluntly that there was no point in dragging along, that unless funds were appropriated for a real airship program the Army might as well close up shop. And this step Congress, in the end, took, and the Army blimps and equipment were transferred to the Navy, and the experimental program started by the one service was carried on by the other.

The rigid ships were in more favorable position because they seemed to have commercial possibilities, and it was the long-range policy of the government to aid transportation. Government support to commercial airships could be justified under the policy by which the government gave land grants to the railways, built highways for the automobile, deepened harbors and built lighthouses for the steamships, laid out airports for planes, gave airmail contracts to keep the U. S. merchant flag floating on the high seas and air routes open over land.

On this theory Navy airships, even though semi-military, got some support during the reaction period, because they might blaze a trail later for commercial lines—which, with ships and crews and terminals, would be available in emergency as a secondary line of defense, like the merchant marine.

The little non-rigid blimps remained the neglected Cinderellas of post-war days.

The Goodyear Company at Akron, which had built 1000 balloons of all types and 100 airships during and after the war, stepped into the picture during this period with a modest program of its own. The first of the Goodyear fleet, the pioneer, helium-inflated Pilgrim, now in the Smithsonian Institute, was built in 1925.

The Atlantic crossing of the Graf Zeppelin in 1928 and its round-the-world flight in the following year gave new stimulus to all aeronautics. With a relatively tiny Goodyear blimp as escort, the Graf lands at Los Angeles after crossing the Pacific.

At Lakehurst the Graf tries out the "Iron Horse," the U.S. Navy's mobile mooring mast, finds it highly useful, utilized masting equipment thereafter to compile an unusual record for regularity of departures, even under highly unfavorable weather conditions. (U. S. Navy photo)

The U.S.S. Akron, first result growing out of renewed interest in aeronautics after the reaction period, goes on the mast inside the Goodyear air dock, prior to leaving for her trial flights.

No large ground crews are needed with the mobile mast. Even the mighty Akron swings around easily at anchorage, heads into the wind like a weather vane, its control car resting on the ground.

In building this ship, Mr. Litchfield and his company indicated their belief in the value of big airships for trans-oceanic travel, for which the blimps would provide inexpensive training for pilots, and experience in operating under varying weather conditions.

The Pilgrim, the Puritan, the Vigilant, the Mayflower and the rest of the Goodyear fleet which followed—named after cup defenders in international yacht racing—would also uncover during the course of day-after-day operations, improvements in ships and operating technique, which would be available to its customers, the Army and Navy.

In building its own ships, Goodyear was following the tradition of American industry, which does not sit back and merely build goods to order, but has sought by developing better goods to anticipate and stimulate customer demand. In the automobile industry, for example, self-starters, closed cars, steel bodies, balloon tires, streamlining, and the rest were initiated by industry to increase public acceptance and further popularize the automobile. By building its own airships and flying them, Goodyear hoped to expand the market for military and commercial airships.

The doldrum period, which made progress difficult, came to an end with dramatic suddenness. In the year 1927 a youthful pilot flew an airplane, alone, across the Atlantic ocean, and in the following year a middle-aged scientist made a round trip from Europe to America by airship, with 24 people aboard. The imagination of America and the world took fire. Aeronautics started anew.

Perhaps no events in years have appealed so fully to the public consciousness or had such dynamic effects. Almost from the day of Lindbergh's flight and the Graf Zeppelin's arrival at Lakehurst, aeronautical engineers found themselves with money to spend in research and machinery. Airports unrolled across the carpet of America, night lighting came in, pilots became business men, appropriations were rushed through Congress, state assemblies, and city councils, and aeronautics became Big Business almost over night. The period of inaction and of reaction was over.

CHAPTER VI
AIRSHIP IMPROVEMENTS BETWEEN WARS

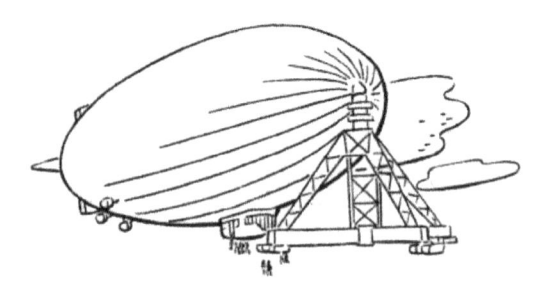

The wartime airship was a cigar-shaped gas bag with an airplane cockpit, open to the weather, slung below. The contrast between it and the sleek, fast, streamlined Navy airship of today is almost as striking as that between wartime planes and automobiles and modern ones.

Many improvements have been made, even though the airship has not had the experience of building thousands of units, as the automobile and airplane have had, or ample funds for research and experiment. Less than 150 non-rigid airships have been built all told since 1914.

The "B" type blimp, chiefly used in the World War, contained 80,000 cubic feet of hydrogen, though some British and French non-rigids were built in larger sizes, and the United States Navy "C" ships, toward the end of the war, had 200,000 cubic feet of lifting gas. These compare with the 416,000 cubic feet of helium in the new Navy "K" ships. Speed, under the pressure of war needs moved up from 47 miles in the "B" to close to 60 in the "C," but is around 80 in today's "K" ships.

Wartime ships carried three to five men and a day's fuel. Today's carry eight or ten, enough pilots, radio men, navigators, riggers and mechanics for two full watches, though normally everyone is on duty during patrols. The "B" was good for perhaps 900 miles, the "K" for well over twice that distance.

Wartime ships had to keep the control car well away from the bag to prevent sparks from igniting the hydrogen gas. A windshield was the pilot's only protection from the elements. Modern ships, using non-inflammable helium, have closed cars, streamlined into the bag, ample room for navigation and radio, sleeping and eating quarters, even a photographic dark room, can be heated and noise-proofed.

Early airships were pulled down and held by a large ground crew, a pneumatic bumper bag on the car cushioning its landing. Today's ships land on a swiveled wheel, roll up to a mast—or taxi off across the airport like an airplane and take off.

These, however, are merely flight factors. More important is it that the wartime blimp was to a large extent hangar-bound. It could go no further from its base than it could safely return before its fuel was exhausted.

Today's ships are expeditionary craft, can go almost anywhere, stay as long as they want. They are no longer land-bound, can be refueled and reserviced at sea. They are much safer, rank high in this respect among all carriers whether on land, sea or in the air.

Three independent lines of study contributed to these results, those of the Army, Navy and Goodyear, each free to follow its own ideas, to observe results found by the others, adopt them, use them as starting points for further developments, or discard them.

The improvements were achieved in a relatively short period. The army started in after the war and carried on a continuing program till 1932. The Navy, absorbed in its rigid airships, did not get into non-rigids till the early 1930's. Goodyear built the Pilgrim in 1925 but its development program really began with the blimp fleet in 1929.

Noteworthy improvement was found during this period in materials, structure, design, engines and radio communication, with outstanding advances along three major lines.

First was increased safety, permitted by helium gas. Wartime airships used hydrogen because it was all they had, had to develop what protection they could against fire through construction devices and operating technique. Hydrogen was not only inflammable, but under certain conditions explosive. World War pilots had to fly their hydrogen ships through thunder and lightning storms, dodge inflammatory bullets if they could. Zeppelin sailors wore felt shoes, with no nails to create a spark, used frogs for buttons, had to guard against static.

It was a fortunate thing for the airship world when a gas was found in 1907 in Dexter, Kansas, which would not burn. Curious scientists, asking why, found it was helium, a gas previously identified (in 1869) only in the rays of the sun. Helium gas is inert, refusing to combine with any other element, does not deteriorate metal or fabric. It was not much heavier than hydrogen, the lightest of all gases, so proved a welcome gift to lighter-than-air.

For some reason, not explained except on the theory that Providence takes special interest in America, helium has been found in quantity only in this country. It is a component, present to the extent of two or three percent in certain natural gas, though ranging as high as eight or ten percent in favored areas. It can be separated by compression and liquefaction from the natural gas,—which is that much improved by the removal of the non-inflammable content.

The world's chief known supply of helium lies in certain sections of Texas, Kansas, Colorado and Utah. More important, United States is the only country having great pipe lines, can distribute natural gas from Texas to cities as far away as Kansas City, St. Louis and Chicago. Without such a market operators would have to separate and release the 95% of natural gas to get the 5% of helium, and costs would be still higher.

Helium is perhaps the most useful of the few natural monopolies given to this country.

It was only toward the end of the World War, however, that Army engineers worked out a process of separating helium from natural gas. A plant was built at Fort Worth and the first cylinders of helium had reached New Orleans ready for shipment to France to inflate observation balloons when the Armistice was signed.

Army, Navy and Bureau of Mine engineers worked thereafter to increase production and cut costs, but as late as 1925 Will Rogers called attention to the fact that the Navy had not been able to get enough helium to supply both the Shenandoah and the Los Angeles at the same time. If one was using the helium the other had to stay home. Two ships, and only one set of helium, he commented.

The use of helium cut the casualty list on the Shenandoah, would have saved the Hindenburg. Non-rigid airships have had no fire or explosive accidents since helium came into use as the lifting gas.

It was the loss by a hydrogen fire of the Italian-built Roma, after it struck a high tension line at Langley Field in February, 1922, which fixed the policy of "helium only" for U. S. Army and Navy airships. The Army's C-7 was the first airship to use helium. In building the Pilgrim in 1925, Goodyear followed the same policy—even though it had to pay $125 a thousand cubic feet for helium while it could have obtained hydrogen for $5 per thousand.

Further improvements and increasing volume of production brought the cost down in time from $125 to less than $20, and helium expense became relatively unimportant in providing safety for Goodyear's airship operations.

Important too during this period was the Army's development of tank cars for transporting helium. A large item of helium expense was freight, the cost of hauling 130 pound metal containers which held 170 to 200 cu. ft. of the gas. It took 250 such containers to inflate Goodyear's smallest ship, the Pilgrim. The tank cars hold 200,000 cu. ft. of gas, almost enough to inflate two Goodyear airships.

Experiments with specially woven fabric and the use of synthetic rubber cut down the losses resulting from diffusion, and where formerly it was necessary to remove the helium and purify it every six months, diffusion losses were cut to one or two per cent a month, with purification needed only every other year.

In addition to increasing safety, helium permitted improvements in airship design. The wartime craft had its control cars suspended by cables from finger patches cemented to the outside of the bag. But with helium ships the car could be built into the bag, attached by an internal catenary suspension system to the top of the gas section. Each exposed suspension cable, no matter how small, creates parasitic resistance from the air, so that the removal of yards of steel and rope had the result of increasing the speed of the ship with the same horsepower.

The second set of major improvements centers around the mooring mast. The mooring mast idea was not new. The British had built the first ones during the World War for its large rigid ships, found that a ship attached to it would swing easily, like a weather vane, continuing to point into the wind, and that a well streamlined ship would hold securely even in winds of great velocity.

When Alfred E. Smith ordered a mooring mast built on top the Empire State building, it was with the assurance from his engineers that even with the tugging of the 150-ton Graf Zeppelin, the strain would be little more than the normal push of the wind against the building itself, that the added stresses would be negligible.

The Germans had had little occasion to use mooring masts. Friedrichshafen, where most of the Zeppelins were built, lay in a natural bowl, well protected from the winds, and ships could take off and land, be walked in or out of the hangar with little risk from the weather.

Lakehurst, on the other hand, lay in an exposed position, in the path of coast-wise storms, a frequent battle-ground between onshore winds from the ocean and storms breaking over the mountains from the west. A study made later to determine bases for projected American passenger operations showed that of weather conditions prevailing between Boston and the Virginia Cape, those at Lakehurst were almost the most unfavorable.

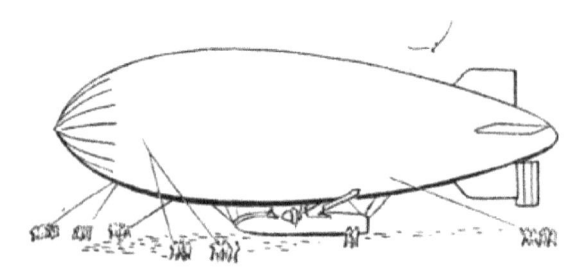

Four stages in the evolution of the mooring mast. At the outset large ground crews held the ship on the ground.

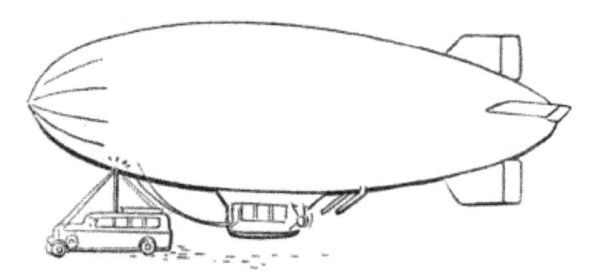

Then a stub mast was placed atop a truck, to hold the ship on the ground, maneuver it in or out of the dock.

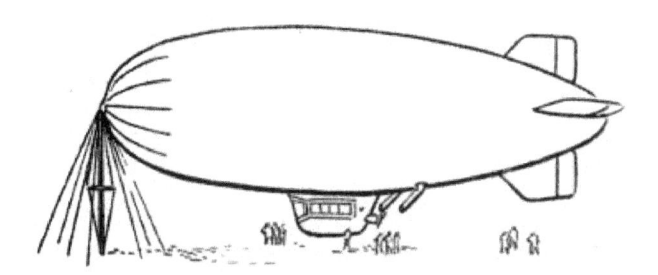

A high mast, made in sections, can be erected anywhere, anchored by guy wires, holds the airship securely against winds of gale force.

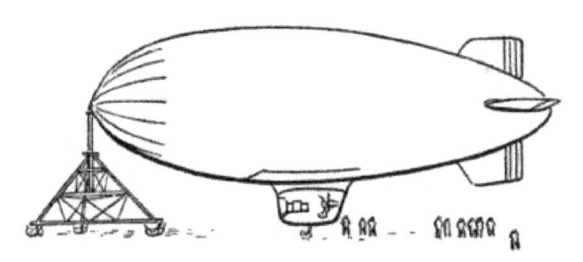

The little brother of the "Iron Horse", which will receive the largest of the new Navy blimps, maneuver them on the field.

People knew little about airship operating when the Navy base was moved from Pensacola to Lakehurst on a waste site in the Jersey pine lands which the Army no longer needed after the war as a proving ground for its artillery.

This defect proved an advantage. The Navy was forced by the very nature of things to concentrate on a problem which had been no problem to Doctor Eckener and his associates. At the urging of Admiral Moffett, Commander Garland Fulton, Lieutenant Commander C. E. Rosendahl and others, Navy engineers built a high mast, 180 feet tall, following British practice, with a service elevator inside, then tackled the problem of keeping the ship on even keel against up and down gusts. Since the wind does not come out of the ground, a low mast was suggested, half the height of the ship, so that when anchored the ship would all but rest on the ground. The Navy was working on this when an incident happened to strengthen the argument.

The co-incidence of a wind shift, and rising temperatures one afternoon as the Los Angeles was resting comfortably at anchorage, started the tail rising, and it continued to rise till it reached almost 90 degrees. Then the ship turned gently on its swivel, and descended easily on the other side, with no more damage than some broken china in the galley. Still a 700-foot airship has no business doing head-stands, so the low mast development was rushed through. It proved successful.

The next step was to make the low mast mobile, so that it could not only hold the ship on the ground but take it in and out of the hangar. First of these was Lakehurst's famous "iron horse," a giant motor-driven tripod, which rolled out on the airport, hauling incoming ships into the hangar, took advantage of daylight calms to take ships out into the field ahead of time so as to be ready to leave on schedule.

On the Graf Zeppelin's trip around the world in 1929, hangars were available for fueling stops at Lakehurst, Friedrichshafen, and curiously enough in Japan, a German shed turned over to the Nipponese after the 1918 Armistice, having been re-erected at Tokio. There was none however on the American West Coast to house the ship after its long trip across the Pacific. So the Navy, under direction of Lieutenant Commander T. G. W. Settle, hauled a mast up to Los Angeles from San Diego (it had been erected there for the Shenandoah's flight around the rim of the country in 1923) anchored it with guy wires. It served the purpose perfectly.

The Germans, skeptical at first, became convinced of the value of the mast, themselves erected masts at Frankfort, and Seville, at Pernambuco and Rio de Janiero, used them as terminals.

Once the masting technique had been worked out, the Graf Zeppelin and the Hindenburg, in the years 1930-6, made a record of regularity which no other vehicle of transportation has approached. They took off at times over the ocean for Europe when all other aircraft in the area was grounded, when the fog hid the entire top half of the ship, and the ship disappeared into the fog within a few seconds after the "Up Ship" signal was given. What few delays appear on the record were due to waiting for connecting airplanes to arrive with the latest European mail for the Americas.

So far the use of masts had been entirely a matter for the large rigid airships. The Army did the first development work on high and low masts for its smaller ships at Scott Field, as well as a landing wheel for them to ride on. A situation at Akron started experimentation along a different line. At Goodyear's Wingfoot Lake Field, Mr. Litchfield frowned over the expense of having a considerable crew on hand to land and launch the blimps, with little to do after the ship was in the air. To an Army or Navy post, with plenty of men in training, this surplus of men was no difficulty, but any private corporation operating passenger airship lines would find the expense burdensome.

The Navy L-2, one of the first ships under the expanded program, lands at Wingfoot Lake, Akron, is walked to the mooring mast.

Close-up view of engine and cowling, and swiveled landing wheel.

With a drogue or sea anchor to hold the airship steady, supplies or personnel may be taken aboard at sea. (U. S. Navy photo)

A newly-hatched airship breaks its shell at Akron, will try its wings then join the Navy.

He put the question to his men in 1930, offering cash prizes for the best solution. Out of many ideas, one clear-cut line of progress appeared. This was to make the ground crew truck a maneuvering base, with a mast on top, which could be folded down when not in use. The truck then could not only hold the ship on the ground, but guide it in and out of the hangar with more security than by using a large number of men. Extra wheels mounted on outriggers kept the truck from being turned over by side gusts. In succeeding years the ground crew truck became a traveling mooring point which could follow the ship across country, give it anchorage when night fell, and at the same time act as a traveling supply depot, machine shop, radio cabin, and crew quarters.

A portable mast, built in sections, high enough for ships to mast at the nose, was the next step. It could be set up on an hour's notice, anchored by guy wires and screw stakes for more extended operations. Gradually the airship became independent of the hangar, came to use it only for overhaul and the purification of its helium gas. The blimp could be fueled and serviced completely in the open.

Lacking a dock in San Francisco, at the time of the Exposition in 1939, the Goodyear blimp Volunteer moved up from Los Angeles, based on a mast for five months. The only time it sought shelter was when a splinter from the propeller pierced the bag, causing a leak. The ship flew 60 miles down the bay to the Navy base at Sunnyvale, like a boy coming in from play to have a splinter removed from his finger, went back again, didn't even stay over night.

In the winter of 1940-41 the "Reliance" which had been spending its winters in Miami, using a wartime Navy hangar which the city had moved up from Key West, found that building commandeered for defense work. So a mast was set up on the Causeway, and the ship operated with no other home than that for six months, saw no shelter from the time it left Wingfoot Lake in early December till it returned at the end of May.

The Navy had a different problem as it moved into the non-rigid picture in the early 1930's. Its problem was only incidentally to operate away from its base at Lakehurst. Ships were getting larger in size, and masts were needed where they could be moored outdoors, or taken in and out of the hangar. The solution was a smaller replica of the rigid airship's "Iron Horse" except that it moved on large rubber tires, and was towed in and out by tractor, rather than carrying its own power plant.

A portable mast was also developed for the Navy blimps, with a special car to haul it around. This mast could be sent to Parris Island or some point in New England, ahead of time, set up and used as a temporary base for radio calibrating or other missions.

Navy ships basing at Lakehurst have operated for weeks at a time along the coast as far north as Bath, Maine, and as far south as the Carolinas, with a portable mast as headquarters.

Utilization of the mast principle by non-rigid airships not only greatly increased their radius of operation, and cut down landing crews, but increased the number of operating days per month.

Pilots of early airplanes used to go out on the airport, hold up a handkerchief, and if it fluttered, conclude it was too windy to fly. So early airship pilots, with anemometers on the roof of the hangar and at points over the field, judged it too risky to take the ships out if the wind was higher than four or five miles an hour, and then only if it was down-hangar in direction.

Modern airships lose few flying days because it is too windy to go out. Under war conditions, when risks must be taken, which need not be taken for passenger or training flights, very few days would be wasted if there is military necessity for it.

Navy non-rigids miss few rendezvous with the fleet in exercises out of Lakehurst, regardless of the weather outside.

If the portable mast revolutionized airship operations over land, experiments started by the Navy in 1938-39, largely under the direction of Lt. C. S. Rounds, promise to be just as important in over-water operations. These showed that the airship could pick up ballast from the ocean, could get fuel from a passing ship, could change crews at sea.

Ballast is important to a vehicle which growing continuously lighter as it uses up fuel, must still be kept in equilibrium. Transoceanic Zeppelins, using hydrogen, had to fly high enough to "blow off" the surplus gas once or twice during a trip to compensate for the ship growing lighter. But hydrogen was cheap, and could be manufactured as needed. American ships could not afford to waste helium, which was a natural resource. Army and Navy engineers had worked on this, and equipment developed for the Akron and Macon to condense the gases from the burned fuel was able to recover more than 100 pounds of water ballast for every 100 pounds of fuel used.

The blimps didn't use these since they ordinarily would not be out for more than a day at a time, still a ready source of ballast would make it unnecessary to valve helium on long flights.

Ironically enough a whole ocean full of ballast lay below seagoing airships, but no practical method had been devised to take the sea water aboard until the Navy tackled the problem in 1938.

That problem may be visualized in the obvious difficulty of maintaining physical contact between an airship and a surface ship. The two move in different media, one influenced mostly by the waves, the other mostly by the wind. The surface ship is moving up and down, the airship subject to gusts which might break the contact or thrust it violently against the masts or superstructure of the surface ship. Servicing has been done under favorable circumstances, but could not be relied on as standard procedure.

The solution reached was this. The pilot swings his ship down to within 100 or 150 feet of the water, lowers a hose with a small bronze scoop, not much wider than the hose, so as to lessen the drag.

Twenty-five feet up from the scoop is a streamlined cylinder, blimp shaped, carrying a small electric pump. This cylinder, nicknamed the "fish", has tail fins to keep it from spinning, and skims along the surface or jumps out like a porpoise, but the scoop is far enough behind and heavy enough to trail easily beneath the surface, stays directly in the ship's wake, continues without interruption to pick up ballast for the airship above.

The whole gear weighs slightly more than 100 pounds, can pick up water at cruising speed, can function in rough water or smooth. The Navy J-4, chiefly used in these experiments, normally consumes 500 pounds of fuel in five hours of flying at cruising speed. It was able to pick up that much water ballast in seven minutes.

The next step was to enable an airship to obtain fuel from a tanker or other ship without physical contact or advance arrangements—even from a passing merchantman. The pilot asks by radio or voice whether the surface ship can spare some gasoline, and on an affirmative answer, lowers or drops on his deck two rubberized fabric spheres connected to each other by 14 feet of rope—also a note of instructions. The smaller sphere is an ordinary air-filled buoy, the larger, about three feet in diameter when filled, is the fuel bag. The surface ship fills the fuel bag, then drops both bags overboard, being careful only that they do not get tangled up. Then the airship flies over the two bags, drops a hook between them, hauls away, pumps the gasoline into its tanks.

The third device permits an airship to anchor in the open sea near a surface ship to transfer crews or take on fuel and supplies. The anchor is a cone-shaped rubberized fabric bag, ten feet long, with a diameter of 2½ feet at the top. It is lowered 50 feet below the airship by two cables connected with each other by rungs to form a ladder. Half of the cables' length is made up of heavy exerciser cord to dampen the effect of wave movements. On top the cone is a wire mesh cover which allows the water to pass through, and is strong enough to act as a platform, supporting a man.

As the cone fills up the airship drops ballast till its "mooring mast" is half submerged. The principle of the drag rope comes into play—if the airship starts to rise it finds itself lifting an increasingly heavier load, counteracting the rising tendency. If it starts to settle down toward the water, the load is correspondingly lessened and the ship grows lighter. The result is that the airship is held highly stable, even in a rough sea. The surface ship then sends a small boat alongside and dispatches the relief crew members or supplies, them up and down the ladder, or uses a winch, the platform atop the anchor serving as the operating base. This system also permits the moving of a sick passenger ashore, or the rescue of a man overboard.

When the airship is ready to leave its anchorage, the cone is tipped by a line attached to the bottom, spilling the water, and hauled aboard. The servicing ship need carry no special equipment. The weight of cone and ladder is negligible.

By being able to pick up ballast and borrow fuel from a passing ship, (neither airship nor surface ship need slow down for the fuel exchange if going in the same direction) the airship greatly increases its radius of operations.

The advantage of being able to change crews at sea may not be quite as clear. This, however, grows out of the fact that today's non-rigid airship has greater endurance than the crew which flies it. An anti-submarine, anti-mine patrol calls for constant alertness. Reduction of vibration and noise, the use of closed cars instead of open cockpits has lessened fatigue, enabling men to remain on duty over longer periods than before. But obviously there are limits.

The Navy is conservative in estimating how long its new "K" ships may stay out without refueling. Weather and the nature of the mission will have some bearing on that, but if we assume a cruise of 48, 60 or even 72 hours which might be done under favorable conditions and idling the motors, we still cannot expect a crew of men to remain vigilant and alert for that length of time.

Extra men for relief watches can be carried only at the expense of the fuel load. However, if a fresh crew could be sent aboard every 12 hours from a nearby surface ship, along with fuel, ballast and supplies, the blimps might operate for extended periods.

No blimps have done this. The fleet might see no need for them to go out for long periods. However, the possibility has been established, and might be useful in the emergencies of war, or accident. While the primary usefulness of the blimp lies in the coastal waters, it can go to sea if needed—and stay out—can be used in convoy work or as a listening post.

Other improvements were uncovered during the experiments. A sea anchor or drogue was devised to enable the airship to "lay to" for extended periods, without consuming fuel, in case it wishes to use its listening devices against submarines, make repairs or for other purposes. Plans have been worked out for landing on the water in quiet bays in calm weather, utilizing flotation gear, or a three-point mooring to ordinary mud anchors—facilitating servicing from nearby Coast Guard stations.

Perhaps a significant thing about these experiments is that the principles seem applicable as well to rigid airships. The ability to pick up ballast in flight may well eliminate the necessity for ballast-recovery devices, with a substantial saving in cost, and an impressive saving in weight.

By eliminating the heavy condensers, and translating that weight-saving into fuel, it is estimated that the range of a ship of the Los Angeles size could be increased by 20 percent and ships of the Akron-Macon size by 15 percent, in the last case amounting to 1,250 miles of additional cruising radius.

A trans-oceanic passenger airship could start out with virtually no water ballast at all except a minimum amount for maneuvering, use its fuel supply as ballast and pick up sea water as needed. This could be done at 500 feet elevation, at the rate of 80 gallons a minute, using a 30 horsepower motor, could be done in half an hour a day. The ship need not slow down materially while doing this.

Application of this principle to military airships of the rigid type might be still more significant. The chief use for the rigid airship in war would seem to be as a high speed airplane carrier, whose planes would increase many fold its own reconnaissance range, and would be expected also to do the major part of what fighting became necessary in case of enemy contact. The airship itself in that situation would put more dependence on its speed of retreat and its ability to seek cover in clouds as the submarine does beneath the surface, than on its own machine guns and cannons.

One thing brought urgently home to us in the first weeks of the present war is that oceans are wide, and that the movements of even a huge enemy fleet are difficult to discover in those endless expanses of water.

Large military airships of five or ten million cubic feet helium capacity might prove exceedingly useful, if they were able to operate away from their base for weeks or even months at a time, and they might be able to do this by utilizing devices similar to those developed for smaller non-rigids, resting on the sea in calm waters, mooring to anchored masts they could lower into the water, picking up fuel from tankers, getting supplies from neighboring ships—in addition to what was carried to them from the fleet by their own planes.

CHAPTER VII
ADVENTURES OF THE GOODYEAR FLEET

One of the lesser romances at least of aeronautics is the story of the Goodyear airship fleet.

There is thrill and adventure in the narrative, daring and resourcefulness, hazards faced by men who believed in their craft—chances which were usually won. So this chapter might well be dedicated to Airship Captain Charles Brannigan and Balloon Pilot Walter Morton.

Morton was an old timer, who had flown balloons with Tom Baldwin, in the far corners of the country. Between times he worked in the Goodyear balloon room, a practical mechanic who could always make things work, the salt-of-the-earth workman whom every foreman swore by, the aide every pilot wanted alongside. Steady, self-effacing, courageous, with an instinct for the right thing to do in emergency, Morton feared but one thing. That was lightning.

He had flown many times through lightning storms prior to the helium era, beneath a bag filled with inflammable gas, but he didn't like it. He knew its swift striking power.

"I could almost see the Old Fellow standing there throwing those darts at us," said Morton one afternoon in 1928, as he scanned the skies before taking off in a balloon race out of Pittsburgh. "One would flash past and miss, and he would say

'I'll get you next time,' and there would come another. And you can't dodge in a balloon."

The Old Fellow scored a direct hit that afternoon. Morton was flying with Van Orman, Gordon Bennett Cup winner. The uncertain weather of the afternoon had resolved itself less than an hour after the take-off, and eight balloons were being tossed as a juggler tosses weights, a thousand feet high, 10,000 feet, caught and tossed aloft again just before they touched the ground. Morton's balloon was hit at 12,000 feet, caught fire, alternatively fell like a plumb bob or parachuted in the net, landed without too much of a shock. Van Orman, unconscious, sustained a broken ankle. Morton had been instantly killed.

But aerologists learned things that afternoon about the force of vertical movements of the air. The balloons gave a perfect track of what went on. One balloon was falling so fast that sacks of ballast thrown overboard lagged behind it, while a hundred yards away another balloon was shooting upward at similar speed.

We still know less than we should about the movements of the air, this new world into which the Aeronautic Age is moving. The Pittsburgh tragedy may save many lives, avoid other tragedies.

The Brannigan story is shorter, no less dramatic. High-spirited, keen, a captain whose ship and crew must always be shipshape, Brannigan had come to Goodyear from the Army—where he had already distinguished himself by making repairs in mid air to the semi-rigid Roma, ripped by a splintered propeller—saving a comrade as an incident to the job—had quickly won his captaincy at Goodyear, was one of its best flyers.

At Kansas City one afternoon in 1931 a Kansas twister headed for the airport. Seeing the weather uncertain Brannigan had stopped passenger flying, put his ship on the mast. Now he ordered his mechanic to get off and cut the ship loose. Once aloft, with helium gas, he was not afraid of any storm that blew. But before the ship could clear the mast, the storm had struck, with full fury. The anchors holding the mast pulled out of the ground and the ship, with the mast attached, was hurled into the nearest hangar, ripping one motor off. That was Brannigan's cue to jump. The door had been propped open for a photographer's camera. But he had one motor left, the bag was undamaged, the mast had fallen clear. He wouldn't give up his ship as long as there was a chance to save it.

Reunion in Akron—The ships comprising the Goodyear fleet, could tell stirring stories of battles with the elements waged in many states.

Some of these pilots flew airships in the first war, others came in later from the technical schools—many now are flying airships for the Navy.

From this pocket handkerchief size airport, off the Century of Progress Exposition in Chicago, Goodyear ships carried thousands of passengers, from all over America.

The Mayflower landed on the deck of the SS Bremen, took off passenger P. W. Litchfield.

The Enterprise lands to rescue the crew of an ice-locked steamer in Chesapeake Bay.

However the storm was not to be denied, and before he could get altitude, the wind threw the ship into a nest of high-tension wires, set it afire. Brannigan climbed out, walked to a nearby automobile, transferred to a second car enroute to the hospital after a collision—and died the next day from third-degree burns.

He called Furculow, his co-pilot, just before the end, told him to see that the men in the crew were taken care of, that they were not penalized for the loss of the ship. Furculow, now flying airships for the Navy, is not the only man in Goodyear who will not forget Charley Brannigan. It is on such men that the traditions of the service are built. Any cause for which men give their lives cannot be held lightly.

The Goodyear Company had built a few airships of its own prior to the 1925 Pilgrim, when helium became available. Best known of these was the "Pony Blimp" which operated out of Los Angeles from 1919 to 1923, flew passengers to Catalina, worked for the movies in Arizona and Wyoming.

But the real beginning came with the Pilgrim, the larger Puritan and still larger Defender, as the Goodyear fleet came into existence in 1928-29.

Early pilots had no specific instructions except to take the ships out and fly them—fly them hard, find out all they could about them, see what weaknesses and shortcomings there were and how to improve them. It was another test fleet, repeating the history of the automobile.

The pilots were supposed not to get hurt, but they were to fly in all kinds of weather they felt it safe to fly in. They might lose a few ships, but were expected to be able to walk away from them, not to get in any trouble they couldn't get out of. They had an advantage over Army and Navy fliers in having a free hand as to where they might go. They were expected to make mistakes but should learn from them.

Such instructions, largely unwritten, acted as a challenge to the pilots, a high-spirited and courageous group. Starting with a few men who had flown airships in the World War, or helped build them in the balloon room and the machine shop, they added some technical school graduates in 1929, and others as needed.

Their adventures started after they left Akron. Operating from bases built or leased over the country, they would cover every state east of the Mississippi in a few years. They looked for hard things to do—or unusual things which would interest the public in airships. They landed on the roofs of buildings in Akron and in Washington—though a prudent Department of Commerce would later rule against that; they picked up mail from lines dropped on decks of incoming ships, and from small boats alongside; they fished for sharks and barracuda, hunted for whales; they picked up a bundle of newspapers from the Hearst building downtown, and lowered them to Al Smith on the top deck of the Empire State building; picked up another batch from The Toronto Star offices, delivered them at the Canadian Exposition grounds; they covered boat races, football and baseball games, the International Yacht Races, carrying press photographers, newsreel men and radio announcers; they went to the Mardi Gras, to the Carnival of States, the Cotton Carnival, Expositions at Chicago, Dallas, Cleveland, San Francisco and New York, to county fairs, plowing and corn-husking contests. They covered fires in New York, chased outlaws and reported forest fires in the high Sierras; they made traffic studies in New York and Washington, studies in bird life in Florida; they picked up stranded fishermen in the Gulf of Mexico, took Mr. Litchfield off the after deck of the SS Bremen in New York harbor; they surveyed canal projects; patrolled the Mississippi during flood time to rescue families from raging waters, to report to the engineers where the levees were weakening; they carried food and supplies to a boat ice-bound in Chesapeake Bay; they circled a thousand country school houses, dropped greetings by parachute to hundreds of cities.

One of their spectacular feats was the rescue of an airplane crew in Florida in 1933. Two pilots flying to Miami from Tampa for the Air Races had made a forced landing in the Everglades. Searching airplanes located the ship, but it was far from any highway, inaccessible by boat or on foot, the men without food and tormented by mosquitos, and with apparently no way of ever getting out unless a road could be built in to them. But a blimp found it easy, because it alone of all craft could stand virtually still in the air.

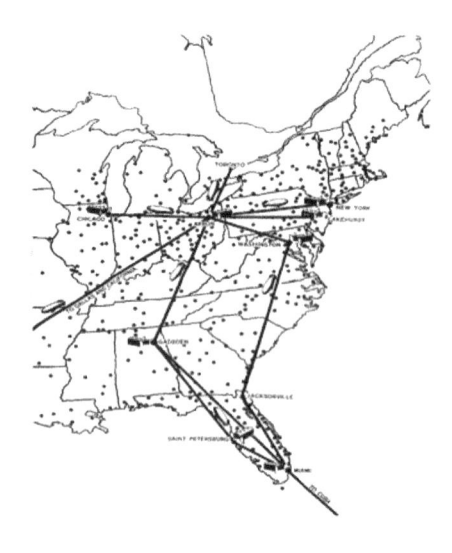

Few important cities east of the Mississippi have missed seeing a Goodyear blimp by now, not to speak of those in the Southwest, the Pacific coast. Trips have been made also to Cuba, Canada and Mexico. More than 400,000 passengers have been carried, without even the scratch of a finger.

SUMMARY	
TOTALS UP TO JANUARY 1, 1942	
FLIGHTS	151,810
HOURS	92,966
PASSENGERS	405,526
MILES	4,183,470
FLIGHTS BETWEEN:	
AKRON - FLORIDA	49
" - DALLAS	6
" - CHICAGO	12
" - TORONTO	14
" - LAKEHURST	18
" - WASHINGTON	57
" - NEW YORK	42

Pilot Wilson flew to the spot, cut his motors, drifted down to 50 feet, directed the refugees to catch the trail ropes, then as the airship settled took them aboard, dropped sand bags to lighten ship, flew home—came back later with salvage parties to recover motors and other parts.

All these exploits were incidental to the job of learning about airships and airship weather—the tricks of winds and rain and storms. And they did learn. A hangar had been built in the woods at Grosse Ile, Detroit, with a lane of trees left standing so as to extend the line of the building—this under the assumption that the trees would protect the airships while entering or leaving. The British, under stress of war conditions had done this, used woods as windbreaks for landings, even for the assembly of airships at times.

But the wind has a trick of spilling over, like a waterfall, when it strikes an obstruction. Early pilots were expert balloonists, and might have remembered their experience in riding over mountainous country—observed how the wind would carry them almost into a cliff, but just before reaching it would pick the great bag gently up, carry it over the top, drop it on the far side, almost to the bottom of the next valley—but not quite, pick it up and carry on—a graphic chart of the air flow in broken terrain.

But in the first weeks of operation at Detroit, a cross-hangar wind, spilling over the windbreak, twice pushed an airship gently but firmly into the trees on the far side. The trees were cut down, and the study of eddies and gusts hastened the development of a mobile mooring mast which would hold the ship steady in turbulent areas.

The Goodyear pilots learned to fly unworried through fog. As early as 1920, Hockensmith, flying the "Pony Blimp" from Los Angeles to Catalina Island, got lost when his compass failed in a fog so dense he could hardly see the nose of the ship. Flying low and slowly, barely off the water, he presently spied a dark shape ahead, came on a U. S. submarine, with decks awash, and an officer on lookout in the conning tower. He landed on his pontoons, taxied alongside, borrowed a compass, went on to his destination.

The conviction that except within its hangar the ship was safest in the air, grew out of many battles with wind and storm. Brannigan, flying the Vigilant at Washington, was caught in a storm which broke up an aeronautic show, wrecked several planes on the ground, sent the rest scattering for shelter. Piling extra cans of gasoline aboard, Brannigan cut his ship loose, headed into the wind, a wind so high that at times he found himself pushed backward at full throttle, hovered for an hour and a half over the capital, waiting the storm out, then flew 150 miles down the bay to Langley field and put up for the night.

On another occasion at Winston Salem, with his ship on the mast, Brannigan was caught in a sleet storm, found his ship bowed down and being crushed by the weight of ice on its back. Getting extra men from the city fire department, he braced his control surfaces with poles, beat off the ice on the bag as high as he could reach with branches, built oil smudge fires alongside to melt the ice, took off all possible equipment, to lighten ship, kept his craft headed into the wind, fought the storm successfully—and in the morning as the sun came out and the ice melted, flew on to Florida.

Boettner, starting south in 1930 in the larger Defender attempting a non-stop flight to Miami, ran into ice and snow in the Tennessee mountains. An oil line froze. His mechanic climbed out on the outriggers and made emergency repairs in flight, but not before the ship had lost most of its oil. Reaching Knoxville airport by morning, he dropped a note, lowered a line, hauled up additional oil, refilled the tanks, went on to the Gadsden hangar to complete repairs.

No Goodyear blimp has ever been damaged by storms while in the air, though a bit of resourcefulness was needed from time to time. For that matter, inquiry does not disclose any cases of a non-rigid airship being damaged by storm while in flight.

Two Goodyear blimps were in the path of the 1938 hurricane, which, heading for Florida from the Caribbean, changed its course erratically and moved up the coast, shot across New England. Lange, with the Enterprise, was at New Brunswick, N.J., 50 miles off the direct course of the hurricane. He put his ship on the mast, held it there during winds which rose as high as 73 miles per hour. He put extra men on the handling lines, doubled the number of screw stakes which held the mast, used the bus, with its motor wide open, as further re-enforcement. The storm raged furiously at the ship for hours but couldn't budge it and when the hurricane passed on, everything was intact.

Boettner, with the Puritan at Springfield, Mass., was almost directly at the axis of the storm. He made the same gallant fight as Lange, but against winds which roared to 100 miles per hour in gusts, uprooted 100-year-old trees, tugged at a sheet-iron hangar roof, flapping it up and down, finally ripped it loose, sailed it like a child's kite across the airport and out of sight.

At the peak of the storm the steel chains attaching the mast cables to the screw stakes failed on the windward side, thrusting the mast into the side of the ship, cutting a hole in the fabric. Boettner pulled out the rip panel, deflating the ship to prevent further damage and when the storm passed rolled up the bag, loaded it and the control car aboard a truck, shipped it into Akron where a new bag was attached. The Puritan was back at work within a week.

No wonder Goodyear pilots came to have great faith in the staunchness of their craft, and their ability to get out of trouble.

Fuel exhaustion didn't bother the blimp. Fickes found that out early, at Wingfoot Lake, when a leak developed in his tank and emptied it. Free ballooning his ship he floated over a farm house, asked them to call the office, waited aloft till a truck came out with additional fuel.

Boettner had a similar difficulty while returning from Canada in the Defender. Persistent headwinds cut down his fuel and when he reached the American shore around midnight it was a question whether he could go on as far as Akron. Picking up U. S. Highway Five as being heavily traveled, he swung low over an adjoining field, slowed down so that his mechanic could drop off, flag a passing car and go into town for gas. By the time the aide returned a number of cars had parked alongside. Driving into the field, with headlights full on they formed a half circle, and the drivers caught the lines, held the ship till the fuel could be delivered, and Boettner proceeded on to Wingfoot Lake.

Mishaps there were of course, in all these years, but few were serious. Lange snagged a lone dead tree in the fog over the Alabama mountains and Smith side-swiped another while flying over a pass in Tennessee. The ship settled easily to the ground in each instance, and farmers came in with stone boats, carried the car and bag to town for repairs.

Brannigan, returning at night from Syracuse, ran short of gasoline, directed his ground crew to land him in an open field ahead. The ship nosed down, his aide directing the men with his flashlight. But just at this juncture the top of the flashlight fell off into the propeller, was whipped into the bag like a bullet, started a leak which was not discovered till next day.

Most ships in the Goodyear fleet have been fired on by thoughtless hunters. Once a bullet went through a ship a few inches back of the pilot. One marksman was arrested and sent to jail in Florida. Pilot Trotter had a curious experience in Oklahoma in 1935, while on his way to the Dallas fair. The ship had been on the mast for three days waiting for weather. On the fourth morning, finding the ship rather sluggish, Trotter looked around. A glass window from the cabin gives a view of the interior of the bag and as Trotter looked he saw light blinking from 14 bullet holes—through which gas had been pouring for three days!

The nearest hangar where repairs could be made and helium secured was at Scott Field, near St. Louis, 400 miles away. By this time the ship had barely enough lift for the pilot and 100 gallons of gas, not enough for the co-pilot. So Trotter flew alone to St. Louis, landing so heavy that the ship had almost to be carried into the hangar, made his repairs and was back in Oklahoma the next day.

Sewell had the experience of seeing a propeller fly off while heading down the bay from San Francisco, saw it careen wildly down, flew on to the next airport on one motor, mounted his spare.

Always the pilots were calling for more speed, removing or streamlining whatever sources of resistance they could, picking the time for cross-country flights when conditions were favorable. They flew from Akron to Washington and New York frequently at 60 miles per hour. The Reliance did even better in a trip north in 1939.

Starting home after its winter in Florida, the ship was held up in Jacksonville—by tire trouble of all things. The distance an airship can make in a day is limited by the distance the bus can travel, since the ground crew must be on hand at night to land the ship. And by now the bus, with its radio equipment, masts and the like had reached the point where only the special Goodyear YKL tires would sustain the 14,000 pounds of weight comfortably. There was a shortage of YKL's when they started and three standard tires had failed on the run up from Miami. Neither Jacksonville nor Atlanta branch had YKL's in that size and to get them from Akron would entail a day's delay.

Meanwhile the ship was tugging on the mast, with a strong south wind, anxious to get under way. The pilots held a conference. Maybe, utilizing the tail wind, they could make it non-stop all the way to Washington, 700 miles north and have Lange's crew land them. If they ran short of gas they could stop at Ft. Bragg, N. C., a convenient half-way point. The Army had a motorized observation balloon there, and was always willing to lend a hand to fellow airshippers. It was Sheppard's turn to take the controls. He sent a wire to Ft. Bragg.

"If I run short of fuel, I'll circle the field as a signal. Could you land my ship, lend me enough gas to get on to Washington?" The answer came back promptly, in the affirmative, and the ship left at midnight.

Roaring across the Carolinas at mile a minute speed the Reliance sighted Ft. Bragg before daylight, with plenty of gas left. An entire company was lined up ready to land the ship. Sheppard flew low, cut his motors, thanked them, flew on for Hoover Airport, arriving before noon. He averaged 66 miles per hour over the 700 mile trip, and landed with enough gasoline to have gone on to New York.

By utilizing helping winds, throttling his motors to cruising speed, Sheppard had effected most economical use of his fuel supply.

Fickes used the same technique more strikingly in the delivery flight of the larger Navy K-5 in 1941, when he flew in to Lakehurst from Wingfoot Lake at 100 miles per hour speed, again demonstrating that greater cruising radius than that for which a ship was designed may be effected, whenever it is possible to pick departure times that are most favorable.

Ships like these, off New York City's great harbor, might afford warning of the approach of enemy submarines, or the laying of mines to endanger its shipping.

Operating from a base across in Jersey, the blimps became a familiar sight around New York City during the World's Fair.

While throughout the middle west, the long afternoon shadows marked the arrival in one city after another of strange visitors from the sky.

Other improvements in construction or operating technique grew out of the fleet's experiences in flying in all weathers. A trip made by the Defender in 1930 from Miami across to Havana brought home the usefulness of the radio. The insurance underwriters insisted on a two-way radio being installed, along with pontoons on the ship, as safety precautions. Neither radio nor pontoons were needed during the crossing, but the pilots sensed the desirability of being able to communicate with their home station and their airport objective. Shortly after a short wave frequency was granted to the ships, one of the early ones in aircraft, and two-way sets were later installed on every ship, on the ground-crew buses and at Akron.

This permitted the making of daily weather maps, extended the airships' radius of action. Pilots would set out with more assurance, knowing that they would be quickly advised of foul weather ahead, could change their course, give appropriate instructions to the men on the ground, land whenever it seemed desirable.

In the end the airships were all doing instrument flying, riding the radio beams like the passenger airplanes, got their landing and take-off instructions from the radio control towers at the airports.

The fleet proved an ideal testing vehicle for the expeditionary mast. But progress moved carefully, a step at a time. As late as 1930 an air dock was built alongside the company's plant at Gadsden, Ala., for use as an operating base in the middle south. It was thought necessary as a half way point for ships headed for Florida. After the high mast came in however, the Gadsden dock came to be used only for warehousing, and no airship has been inside it in four years.

In 1932 the Volunteer started in from Los Angeles for Akron, making the first successful trip of any non-rigid airship over the Continental Divide. The Volunteer was due for helium purification and a new bag. No helium facilities were available closer than Akron. Rather than deflate the ship and send it by train, Pilot Smith decided to fly in. He laid out a route via El Paso, San Antonio, and Scott Field, so that he could get shelter, if necessary, at army hangars at those points. He berthed at El Paso just after a 100-mile-an-hour storm had passed over, stayed three days at Kelly Field, found it unnecessary to stop over night at Scott. Even so, because of persistent head winds he had had to spend ten nights in the open, setting up his low mast with screw stakes on the open prairie.

Mooring out procedure had improved by the time that Sewell made the same trip five years later, so he made only courtesy stops at the three army camps, was on his own.

A mishap at Louisville gave impetus to the development of the high mast. The retractible low mast mounted on top of the bus was attached to the bag about half way between the car and nose of the ship, convenient to get at, the system being referred to as "belly-mooring." The low mast was light, could be set up quickly and easily, would hold securely against a straight pull of considerable force. However, it was not as effective in the case of a wind shift, or gusts which rolled the ship on its side. A higher mast, with the ship anchored at the nose, was free to swing in all directions. Every one realized this, but it was only after Crum's ship was caught and twisted by a gust at Louisville, punching a hole in the bag, that the change was made.

The high mast, built in sections, anchored by guy wires to stakes screwed in the ground, was more bulky, took longer to set up, but would hold the ship indefinitely once it was in place.

Thereafter both masts were carried in cross-country trips, the convenient low mast being used for overnight stops in good weather, the high mast for more extended operations, or when the weather looked threatening.

The ground-crew bus was in evolution during this period. Built originally to carry merely crew, spare parts and supplies it added a radio room, navigation quarters, and carried the two masts. A scout car cruises ahead to make overnight arrangements, a trailer follows, with its own electric plant and expeditionary equipment, including a spot light to play on the ship at night. Duties of airship personnel grew more specialized and complex.

Members of the ground crew acted as radio technicians, meteorologists, mechanics, riggers. They comprised a colorful group, recruited from all parts of the country. Sailors from New Bedford, fruit growers from Florida, farm boys from Ohio, ranchers from the San Joaquin valley, a mechanic from a Chicago airport, a policeman from the Cleveland fair, all dropped their work and followed the airships. The personnel list was a history of every place an airship had operated.

The work wasn't easy, involved long hours in the cold and rain when storms threatened, picking up mail from their families on the fly in cross-country operations, moving their households from north to south and north again. But the ground-crew men stuck, most of them having ten years' service and more. On cross-country trips a crew of 14, including pilots, is adequate.

The pilot personnel too formed an interesting group. Jack Boettner, chief pilot, veteran of the group, with probably more airship hours than any man in the world, certainly in non-rigid airships, had played all-American football at Washington and Jefferson, been instructor at Wingfoot Lake through the first war, was working in Goodyear's aeronautical sales when the fleet got under way.

As expansion started in 1927 Smith came in from the aero workshop, would remain second in flight hours only to Boettner. Fickes from Akron University, left the Efficiency Dept. to sign up, set up one of the first outside bases, at New Bedford, flew the Mayflower when it picked up Mr. Litchfield from an ocean liner, later became manager of all airship operations. O'Neil from the workshop came on too, in that year, became chief mechanic.

When a base was set up at Los Angeles, Lange, a New Englander who had left Boston University to fly airships in the first war, later flying out of Panama, joined up, was sent to California, later took charge of the Washington base. Sewell, a Kansan with a similar record, having left the state university to fly blimps in coastal patrol in 1918 came in, captained a ship at New York, followed Lange at Los Angeles.

Further expansion came in 1929, when the Puritan, Mayflower, Vigilant and Volunteer and Defender were added to the fleet. Now came Wilson, Purdue footballer, Furculow from West Point and Mt. Union, Hobensack from West Virginia U, Rieker and Crum from Ohio State, the last named becoming engineer officer of the group.

Other practical men came in, from the balloon room and aero shops—Sheppard a Virginian, who later flew all over New England, the Middle West and Texas; Massick, Crosier and Munro; Blair, Army sergeant from Scott Field, came to Goodyear after the semi-rigid RS-1 was finished.

Stacy, another New Englander, left the class room at Massachusetts Tech to sign up. Dixon, born in a lighthouse on Nantucket Island, left a billet as junior officer on a South American liner to fly land ships instead. Trotter, from the Naval Academy, was in engineering work in Florida when a blimp flew over. Lueders came in via the ground crew at Los Angeles.

Many of the Goodyear pilots were commissioned as Reserve officers in the Navy, and Fickes, Boettner, Lange, Sewell, Wilson, Trotter and Furculow each took a year's active duty with the Navy at Lakehurst with rigid ships. More than a score of trips were made by Goodyear pilots across the ocean as student officers aboard the Graf Zeppelin and the Hindenburg, getting post-graduate training.

The breaking up of the pilot organization began as early as 1940, when with war clouds appearing in the East, Trotter, Rieker and Furculow volunteered for active duty with the Navy. By the middle of 1941, Stacy, Smith, Lueders and Dixon had followed them into uniform, were flying Navy airships at Lakehurst.

To fill their places and also furnish material for the already expanding airship Navy, a training class of 19 men was started in late 1940 at Akron and Los Angeles. A six-months' ground school preceded flight training—which started with seven balloon flights.

The training course evolved there was one which grew naturally out of such a situation. Airship piloting had changed from the "seat of the pants" flying of the first war, when veteran Jack Boettner would turn out pilots in six weeks. The ships had become more complex as improvements were made. Helium gas was being used. Navigation by radio and compass was quite different from the "concrete compass flying" of 1916, when pilots followed highways or railroad tracks to keep on course. Instrument flying had come in, and blind flying was part of every student's training, in a closed control car, operating by instrument only. The modern airship pilot had to know his radio beams and the rules of Civil Aeronautics Authority, be able to ride the beam into the airport. In these various details the Goodyear pilots, long-seasoned, had perfected themselves through years of operation, were competent to pass on their secrets to the youngsters coming in.

The student pilot spent his first half dozen hours trying only to keep the ship at constant altitude, not caring where he was going. Then he would fly a given course, follow a zigzag rail fence, or a winding road, not worrying about his altitude. Lesson three was to combine the two, fly at constant altitude over a set course. And after enough hours at this, he'd try to circle a pylon, keeping a specified distance away, while the wind pushed the ship in one direction, then another—now flying up wind, now down, now cross-wind, now quartering, making such changes in course to allow for wind and drift as to maintain a perfect circle—and trying finally to achieve the supreme art of the airshipper, which is to get the feel of the controls and the weather so that he can anticipate drift and sharp drops and rises, move his controls a split second ahead of time, stay on course and altitude.

Airship students got no exemption from Civil Aeronautics Authority by reason of the fact that blimps land more slowly than bombers, took the same physical examination, including eyesight. The training course worked out with the government followed closely that for heavier-than-air pilots, with such changes only as were made necessary by the fact that in one case a static lift was utilized chiefly, and in the other case dynamic lift. There was plenty of need for the students by the time they finished their training.

Over the 16 years during which the fleet operations were carried on ship sizes settled down to 123,000 cu. ft. as a compromise between the 51,000 cu. ft. Pilgrim and the 164,000 cu. ft. Defender. This size ship could carry six passengers with pilot and aide, was easy to handle with a small crew, had adequate cruising radius for the job at hand.

Later ships, the Enterprise, Ranger, Resolute, Reliance and Rainbow, carried on the tradition of honoring the defenders of America's cup in international racing.

While an airplane can land anywhere on an open field, the airship needed at least a minimum of terminal facilities. Many groups co-operated at the outset. St. Petersburg, Florida built a hangar; Miami towed a war-time Navy shed up from Key West; Col. E. H. R. Green built one on his New Bedford estate for use in connection with radio studies being made by Massachusetts Institute of Technology. The company built its own at Gadsden, Los Angeles, Washington, Chicago and New York, calling them air docks rather than hangars.

Unused Army and Navy hangars were borrowed in the early years at Aberdeen, Md., and briefly at Cape May, N. J., Pensacola, Arcadia, Cal. and Chatham, Mass., with Lakehurst, Langley Field, Scott Field and Sunnyvale, Cal., handy as ports of call.

More and more, however, the fleet grew independent of ground aid, became increasingly self-reliant through the use of its masting equipment.

The Goodyear fleet wrote a remarkable safety record in the 16 years. Accidents to airship personnel could be counted on the fingers of one hand, and in the case of the public, 400,000 passengers had been carried up to 1942, for a total of 4,000,000 miles without a scratch of anyone's finger.

CHAPTER VIII
RESULTS OF FLEET OPERATIONS

Goodyear airships made some contribution during the 16 years of fleet operations, to flight and ground handling technique. They also contributed to men's knowledge about weather. For wherever it is flying, an airship, by the very nature of the craft, is continually registering the effects at that point of certain components of weather. And the ships covered a considerable part of the country fairly thoroughly.

The nature and movements of air currents can be studied only incompletely from the ground, for conditions there are merely the result of forces aloft. Only two vehicles leave the ground and use the air as highways. Of these the airship is vastly more responsive to changes in temperatures and barometric pressure than the airplane, because of the lifting gas in its envelope, and somewhat more responsive to changes in wind directions and velocities, because of its greater displacement of air.

Goodyear airships have traveled widely, have seen at first-hand the effects of rain and snow, fog and sleet, wind and whirlwind, thunderhead and lightning storm. More important they have been spectators at the unseen battle waged endlessly between cold fronts and warm ones across the great central plains, continued with renewed vindictiveness through mountain ranges and valleys.

The information brought by these voyagers has not been without value to the men in the airport control towers, who are studying weather phenomena in the effort to make flying safe.

A whole new science of weather interpretation has come in with air transport, and the U. S. Weather Bureau has other duties than advising farmers about planting and harvesting crops. It may be merely coincidence that when a new chief had to be selected for the Weather Bureau a few years ago an airship pilot was selected—Commander F. W. Reichelderfer of the Navy, who had long studied the movement of air masses and their effect on flight.

Army and Navy ships put in more actual flying days per month than Goodyear ships, when on coastal patrol, because once out at sea the service ships were out for all day—and an airship, by picking its time, and using its mast, can always get out and get back.

Goodyear pilots had a different sort of job. They were operating over land, flying 100 passengers a day, at 10 to 15 minute intervals, in one town after another. They might suspend operations when ceilings were low, or winds high, or gusty, not because they couldn't fly under those circumstances, but because flights would be less agreeable, and might be hazardous for their passengers. However, the ships themselves, having no shelter at hand, had to stay out and take it. Their job was to interest the people of America in lighter-than-air, and they had to go wherever people were, regardless of what flying weather might intervene.

So between Navy, Army and Goodyear airships operating over a period of years, it was fairly well demonstrated that there is very little unflyable weather for lighter-than-air craft. That is a conclusion of no small importance.

Winds of gale force may make it prudent for the airship to stay in the hangar or on the mast, and conditions of zero ceiling, zero visibility, which ground other aircraft, would make operations hazardous, especially over mountainous country, but even the most adverse weather conditions would hardly keep the airship at home if an enemy was at large. Any time submarines are operating the airship can be available to seek them out.

Another result emerging from the fact of fleet operations was that flying men and construction men, working together, became a closely knit group. Engineers learned to fly ships, and flyers took their turn in the shops. In building airships for the Navy, at the speed demanded by war conditions, the control cars were built in the shop and the envelopes cut out and fitted and cemented together in the balloon room. But operating men, flyers and ground crew men, mechanics and riggers and maintenance men took over from there, put the ships together—assembled them, tested them out, delivered them to the Navy.

Lessons in streamlining gained from building and flying blimps became useful when barrage balloons came into the picture as a new defense weapon.

The mooring mast made the blimps expeditionary craft, eliminated the need for large ground crews, permitted more flying days per month, increased safety.

Floating Navy blimps and barrage balloons, with their curious star-fish tails, give the service dock something of the appearance of a giant aquarium.

Principal use for the rigid airship in wartime is as an airplane carrier, with half a dozen planes to extend its reconnaissance range and determine the enemy's position.

It was this co-ordination between men in green eye shades, working over the drafting board and wind-tanned pilots, studying gray skies and phosphorescent control boards, which enabled the organization to meet the war emergency of large scale production of non-rigid airships.

There was another by-product result arising from the fact that the company, even in the doldrum days, when there were few orders for ships, had kept its engineers at work on research and its ships flying on experimental missions. It all happened suddenly, a colorful circumstance not often found in the sober humdrum of the business world.

A great plane manufacturer, having more defense work than its crowded shops could handle, looked around for some company with experience in the fabrication of light metal, to whom it could farm out some of the details.

Goodyear Aircraft Corporation, the aeronautic subsidiary, was asked to build tail surfaces for Martin bombers. A curious thing happened. Men whose work had been primarily with airships, rather than airplanes (omitting the quite different field of airplane tires, wheels, and brakes) found themselves on familiar ground when they swung over to heavier-than-air construction.

Here was the same problem of getting maximum strength with minimum weight, of selection and treatment of light alloys, of intricate stress calculations, and a hundred ingenious devices to measure those stresses, enabling designers to turn out a scientifically designed structure. The background was there—not to mention their experience and studies in streamlined design— to reduce resistance, get maximum performance from power plants.

The difference was that in the case of the airship savings in weight mount fast, because of size. The importance of light weight and high strength had come home to airship designers years before.

Their experience was directly applicable to the new field. Other orders came in, from Curtiss, Consolidated, Grumman, and soon the huge plant was humming with the production of parts for fighters and bombers.

Then a four-company arrangement was set up by the government to expand airplane production still further, and after that an order for complete planes. The original plant was now jam-packed with lathes and drills, jigs and presses, and three huge new plants were built alongside and across the road, and Goodyear Aircraft Corporation found itself with thousands of men, building not only airships, but airplanes and airplane parts as well.

Every large company took on new tasks in defense, but in this case Goodyear was able to move quickly, and give unexpected support to the airplane program by reason of its long research in a different field. This result, it is true, grew chiefly out of research in rigid airships, rather than non-rigids, but both played a part in another instance—barrage balloons.

England was using them, might ask this country to supply some. The American government too might have use for them. So, long before there was even any hint of orders, Mr. Litchfield threw a new problem to the engineers at Goodyear Aircraft and the operating men at Wingfoot Lake—the job of designing an efficient barrage balloon. They were not to make Chinese copies of foreign balloons, but draw on their experience in lighter-than-air and see if principles and technique established there could not be applied to design balloons which would ride with maximum stability in gusty and unstable air. Men went to work, designing, building, flying, observing, rejecting, altering, improving, week after week, month after month, until several satisfactory types were evolved. One of these was capable of flying at 15,000 feet, twice the usual height. Orders began to come in, and the little group of men and girls in the balloon room quickly grew into a large organization. The department outgrew its quarters, took over room after room, expanded to subsidiary plants outside Akron.

One instrument developed illustrates how the airship men were able to utilize past experience in a new project.

Mounted alongside the winch on the ground, it gave exact information, as often as was wanted, as to what the barrage balloon was doing, a mile or three miles up.

This assembly included a moving picture camera, which continuously, or at fixed intervals, or at any instant desired, by means of radio control, would photograph recording dials and show these things: wind velocity at the balloon, tension on cable, gas pressure inside the balloon, temperature of confined gas, temperature and humidity of the air surrounding the balloon, angle of attack at which the balloon faced the wind, both fore and aft and from side to side, also a clock, which showed the time the readings were recorded.

These pictures, when developed gave the engineers the data from which they could modify designs and arrive at a type of balloon which would ride most easily aloft, avoid undue tugging and surging on the cable—incidentally permitting smaller gauge and weight cable to be used for a given height with ample safety margin.

Perhaps the largest single result, however, growing out of the fleet operations was that it had created manufacturing facilities, ships and personnel on which the Navy could draw, as fully as it wanted, in emergency, and with little more delay than the time it took for a man to change his uniform.

Boettner, Sewell, Blair, Hobensack and Hill followed the others into the service. Hobensack's ground crew in California signed up with him in a body, and men from other ground crews, expert in rigging, in motors, radio, in mooring out and maintenance joined up. In the end only Fickes and Crum were left at Akron to build the new ships, and Sheppard, Crosier and Massic to test-fly them, then ferry them to their destinations.

The student pilots at Wingfoot Lake had finished their training just in time. About half of them went immediately into the Navy, were commissioned and sent to the various bases, the others remained at Akron as replacements to the other pilots, in testing and delivery flights, or on key posts in airship construction.

The experience accumulated by the blimp pilots under varying weather conditions over the country proved useful to the Navy, particularly in the expeditionary operations which coastal patrol would demand. It was useful as well in helping train navy aviation cadets for the growing airship fleet. Five of the pilots, Sewell, Boettner, Rieker, Stacy and Smith had reached the rank of lieutenant commander by the end of 1942, and Lange, full commander, had become commanding officer of a new Navy station on the west coast. Two of the public relations men, Lieutenants Petrie and Schetter, old airship troupers, followed the fliers into uniform.

The airship service suffered its first casualty in 1942 when Lt. Trotter, gallant and resourceful pilot of balloons and ships, was killed in a collision, in which Lt. Comdr. Rounds also lost his life.

The Goodyear fleet passed out of existence with the war. The ships being the same size as the Navy training ships, it was a simple matter to change them over, paint the new name on their broad sides.

Facilities for ship construction became useful also in the new war. An airship hangar is unlike any other structure in the world. It must be broad and high and free of supporting girders. There were two large airship docks at Akron, half a dozen smaller ones over the country. At hand, too, was equipment for helium purification and storage, along with radio and weather gear, mobile mooring masts and other specialized equipment which only lighter-than-air uses. There was the balloon room, too, with a wealth of experience dating back to the first World War, and which with new jobs like building barrage balloons, rubber rafts and assault boats grew to large dimension.

Wingfoot Lake was more than doubled in size, and the large airship dock, occupied at first by heavier-than-air production, had to be changed back later for airship assembly, to meet the Navy's mounting demands for ships. The bases at Washington and Los Angeles were converted to other aeronautic uses; the two-ship dock at Chicago and the one at New York were torn down and moved to Akron to provide additional space for ship assembly.

And so the fact that the company had maintained an airship fleet for a number of years had the result that in emergency when the Navy needed ships and men to fly them, Goodyear was ready. All of which was not foreseen when Mrs. Litchfield pulled a cord to release a flock of pigeons and christen the pioneer ship Pilgrim, at a pasture-airport outside Akron in 1925.

CHAPTER IX
VULNERABILITY OF AIRSHIPS

Mention airships and most people will immediately raise the question of vulnerability.

Large, slow moving, a tempting target, airships could be shot out of the sky by ship or shore guns, or by hostile airplane fire, it is argued, almost as easily as a dinner guest touching his cigaret to a toy balloon.

And this is probably true, with reservations, if enemy ships or anti-aircraft batteries or planes were around. But the airship, non-rigid, has no more business in such areas than a British airplane carrier would have to drop anchor in Hamburg harbor.

It was because of the imminence of attack from sea or shore or air that neither England nor Germany used airships in the present war, particularly since they would have to use the inflammable hydrogen gas. It was because such attack on American airships from any of these three sources was much less likely—and that we have helium gas, which does not burn—that this country is using them.

Their chief field of operations is not off the enemy's coasts but our own, along that broad ribbon of waters used by our coastwise shipping, an area roughly marked in the Atlantic by the 100 fathom curve, the favorite fishing grounds of enemy submarines. Thousands of miles of blue water, not the narrow lanes of the North Sea or British Channel are between them and the shore guns of an enemy.

An enemy fleet, though likelihood of this seems remote, might penetrate those coast waters in attempted invasion, attack the blimps with anti-aircraft fire. But such an enemy, arriving in force, would have either to knock out our Atlantic fleet, or slip past it in surprise attempt. In the remote later contingency, the information relayed back by airship radio that the enemy was moving in would be worth losing airships or any other craft, to get.

The third hypothesis, attack by airplane, is also conceivable. But if long-ranging enemy planes were able to get that close to our shores they'd have more important business in hand than wasting time and powder on a helium bubble bobbing in the air, 10,000 feet below—which in any event would already have radioed the news ashore.

In the fairly remote contingency that the airplane did choose to attack the blimp, it would find the position of that moving target, flying at an indeterminate distance below, much more difficult to calculate than a fixed target ashore, no easy thing to drop bombs on.

If it swung down close, it might riddle the bag with machine gun bullets but without necessarily sinking it—as witness the case of Trotter's ship in Oklahoma leaking gas for 72 hours from 14 gaping holes and still able to fly 400 miles for repairs. The plane would have almost to cut the blimp in two with a spray of bullets to destroy it—if it chose to use its precious far-borne ammunition in such fashion—and would find it better to attack from below, on the chance of a lucky hit into the airship structure or controls, or one which disabled its crew. But in that event the airship, also armed, shooting it out from its more stable gun platform above would have as good a chance as the plane.

The airship is vulnerable—as are all other military craft—but used as the Navy proposes to use airships, it may be said to have an acceptable degree of vulnerability, in view of its potential usefulness in its special field—defense against submarine attack on convoys or coastwise shipping.

The airship's advantages have been pointed out, but may be repeated. These grow out of its speed range, from zero to a maximum of 65 knots or so. Its slow speed, as compared to the airplane has the compensation that it does not have to circle around to maintain altitude, can keep any suspect object under continuous observation. Its high speed enables it to reach a given point much sooner than the fastest surface scout.

Barrage balloons—spiders who spin out webs of steel as they ascend—but these spiders are out to catch fliers, not flies, enemy fliers who threaten our democracy.

Modern armies towing a few of these pocket sized barrage balloons along, might not be too much concerned over attacks by strafing airplanes.

This Strata Sentinel will fly at 15,000 feet, twice the height of other barrage balloons. By that time the lobes will be completely filled out by expanding pressure of the lifting gas.

This airship, silhouetted against the afternoon sun might be pacing a peaceful cruiser race through the surf off Long Beach, on the Southern California coast. Or it might be leading units of the mosquito fleet to sea off Cape Cod, to hold an enemy U-boat in check till ships of heavier armament could arrive.

Helium-inflated, fast, long ranged, the modern K-type Navy patrol ship is a far cry from the primitive airships of World War I. They are armed with bombs and machine guns.

In brilliant sunshine, or overcast, in fog or rain or snow, the blimps take off from their bases day after day, on guard against any enemy who may invade the coastal waters. A faint smoke screen, miles distant over the endless waters, may turn out to be a peaceful merchantman—or a vessel with grimmer purpose, seeking the advantage of surprise attack.

The detection of a submarine even on the surface is largely a matter of looking in the right direction at the right time. The open windows on all sides of the airship, without obstruction by wings give it special value in this field.

A submarine submerged is still harder to find as its tell-tale feather is not easy to spot from a speeding plane or from the crow's nest of a surface craft.

A non-rigid airship throttling down to the speed of its prey, and having the altitude of the airplane, has a much better chance of sighting the submarine, before it can launch its torpedoes.

Taking off in fog, flying in low visibility, compelled to fly close to the water, these factors do not worry the airship or handicap its usefulness overmuch, and might under given conditions prove extremely useful.

The airship appears to have some advantage too in the length of time it may remain on station, ranging from 30 hours at high speed to undetermined days at low. Indeed its endurance is not so much a matter of fuel capacity as of the ability of crews to stand long watches without relief.

There might be emergencies where airship scouts were wanted on continuous duty over a considerable period. Commander Roands' experiments point out interesting possibilities in this respect, through the transfer of fuel and supplies from a surface ship, and the taking on of fresh crews.

This generally was the case men saw for the airship up to 1941, as having potential usefulness, in the event of war, against attack by sea.

Then came Pearl Harbor, and America's entrance into a new war. German U-boats, larger, faster, more deadly, moved swiftly in to attack, as if waiting for the signal. The Japs made reconnaissance raids along the West Coast.

"Wolf packs" of submarines in new under-water tactics stalked convoys, picked off stragglers. More than 600 coast-wise ships, merchantmen from the Caribbean and South America, and tankers from the Gulf, were sunk in the first year of war. The loss of tankers brought serious complications ashore, the rationing of gas along the eastern seaboard to conserve supply for military purposes. Despite a quickly expanding program of ship construction merchantmen were being sunk faster than they could be built.

The Navy's sea-frontier defense moved to meet the attack. Non-rigid airships were assigned a place in that program, wherever they could be utilized and with what ships were on hand, and new airship construction was rushed.

Under authorization from Congress, a program of airship and base construction, together with helium procurement, was accelerated, and by the end of the year, stations were in commission or being built at key points along both coasts and the Gulf of Mexico.

Akron expanded its facilities many fold for the building of new airships, which were flown to the various bases with increasing frequency during the year. Large classes of officers, aviation cadets and enlisted men went into intensified training at Lakehurst and Moffett Field, preparing themselves to man the ships as fast as they were delivered.

The blimps which have been available to the sea-frontier forces have rendered valuable service in patrol and escort missions. Their exact record of performance, including number of submarine sinkings, obviously cannot now be published.

On sighting a submarine, or finding indication of its presence, the tactical doctrine might call either for attack, or to stand by, summoning airplanes and surface craft in for the kill, keeping the enemy under unsuspected surveillance the while, and saving the blimp's own depth bombs for another action.

The airship is capable of carrying on patrol and escort missions day after day under a wide range of weather conditions, going for months at some stations, even in the winter, without missing a day.

Though no detailed summary of airship activities is possible now, it is no secret that, just as in the last war, the submarines avoided attack upon convoys where airships were on guard. The German high command tacitly admitted that this was one type that the U-boats did not want to meet, an enemy immune to its torpedoes, whose presence the sub's under-water detectors did not reveal, and which might appear overhead without warning. Admiral Doenitz, commanding the German submarine force, testified in a press interview to their respect for our blimps.

The battle against the submarines will be long and difficult, and ships will still go down and men will be lost, but the chase will be relentless as long as the menace exists. Airships, non-rigid, have taken their place in that phase of America's war effort.

REFERENCES

Little is available in the way of bibliography on lighter-than-aircraft, their history and characteristics. Among the best works dealing with this subject are Captain C. E. Rosendahl's, "What About the Airship?" (Scribner's), and "Up Ship" (Dodd Mead); Captain Ernst Lehmann's "Zeppelin" (Longman's) and Captain J. A. Sinclair's "Airships in Peace and War" (Rich & Cowan, London).

Copies of "The Story of the Airship (Non-Rigid)," may be procured through The Goodyear Tire & Rubber Co. at Akron, Ohio; or at Los Angeles, or branch offices.

Milton Keynes UK
Ingram Content Group UK Ltd.
UKHW030846141124
451205UK00005B/450